FIRE

The Spark

That Ignited

Human Evolution

Frances D. Burton

University of New Mexico Press
Albuquerque

15 14 13 12 11 10 09   1 2 3 4 5 6 7

Library of Congress Cataloging-in-Publication Data

Burton, Frances D., 1939–

Fire—the spark that ignited human evolution / Frances D. Burton.

p. cm.

Includes bibliographical references and index.

ISBN 978-0-8263-4646-9 (HARDCOVER : ALK. PAPER)

1. Fire—Social aspects.   2. Fire—History.

3. Anthropology, Prehistoric.   4. Prehistoric peoples.

5. Human evolution.   6. Hearths, Prehistoric.I.   Title.

GN416.B86 2009

306.4—dc22

2009000572

Book and jacket design and type composition by Kathleen Sparkes.

This book is set in Minion OT PRO 10.8/14, 26P.

Display type is Trade Gothic OT LT STD.

To my parents, Emanuel Burton (né Buchenholz) and

Rita Dinah Madeleine Marguerite Burton (née Cahen), who,

like so many displaced persons now too familiar all over the world,

migrated, settled, and made the best of their transplanted lives.

I wish they had lived long enough to see the fruition of the seeds

they sowed; not compensation, surely, for the loss of their own dreams,

but fulfillment nonetheless.

Contents

List of Illustrations

TABLES

FIGURES

Preface

Anthropology is a holistic discipline embracing all sorts of intellectual endeavor in a synthesis. Its major contribution to science—natural or social—lies in the juxtaposition and integration, the bringing together, of what otherwise would be disparate ideas. Geneticists are experts in genetics, anatomists in anatomy, and so it goes for every discipline and its practitioners. Anthropologists cross the boundaries; we fuse notions, producing a "new" perspective or idea. This is the context for my daring to write on the physiological effect of fire in the evolution of humans. My interest in this question first announced itself almost thirty years ago, but the demands of teaching, publication, and family did not support researching something so speculative. Serendipitously, it was just as well that I waited: an explosion of articles on the circadian clock appeared beginning in 2001. Each article added a bit to, or totally refuted, the one before it, and the complexity of the subject seemed to grow exponentially. Not the least of these concerned how humans became human.

There is no doubt that hominization—the process of becoming human—accelerated over time and with reference to other primates. The anthropologist Bronislaw Malinowski described what has since been named "emergent" properties of human culture: a spiraling effect that emanates from basic needs. Ivory chopsticks, for example, have less to do with getting food to the mouth than they serve as a symbol of wealth, class, aesthetics, and even the resources used to make them. Basic needs of the senses, of the body, of the mind are met, satisfied, and exceeded by cultural objects, which in turn give rise to other cultural objects, giving rise to mind, giving rise to body. Much of human evolution reflects this feedback process as described in the following pages.

Acknowledgments

So much effort goes into writing a book that it is hard even to identify everyone that should be acknowledged. For two years before my final sabbatical, students at the University of Toronto Scarborough did research in courses that gave me background and insight into the problem I was exploring. They did exceptional work and received appropriate grades, but my thanks go far beyond academic recompense. They became an inspiration, support, and motivation. In particular I would like to thank Allisha Ali and her exhaustive work on fireplace sites, Corinna Cooke and her pyrophilous insects, Jay Han for his studies on seasonal affective disorder and circadian rhythms, Vromi Fernandez for hers on fossil ancestors, and Jennifer Jairam for her work in the exposition of genetics.

Writing is part of life, and while it may be all-consuming, there are still obligations to be met and pleasures to be had. My daughters gave me the confidence to pursue a topic that is complex and difficult; their children gave me joy and respite from my work. One of the hazards of living in Academe for a long time is subscribing to a particular form of language. My husband, Peter Silverman, intrepid and generous soul, dared to comment on the draft and helped make my use of the English language more approachable. His cajoling, insight, and chapter headings have given the book a bit more pizzazz. Laura Wright edited my first draft, plowing through grammatical errors, misspellings, errors of logic, and other deviations with gentleness and persuasion. I asked, and received, permission from Arnold Chamove to quote his important experiment with chimps and fire, and I thank him not only for the permission, but for the research. The wavelength image was compiled from a scale provided by the National Aeronautics and Space Administration. Figure 2, the eye, is based on an image from the John Moran Eye Center, University of Utah, kindly permitted by Helga Kolb. Pamela Heidi Douglas kindly sent me the picture of a bonobo from Lola y Bonobo Sanctuary in Kinshasa, Congo. This image of a bonobo crushing nuts with a stone tells a story very much in line with the themes of this book, and I thank Heidi for embellishing it with this generous contribution. A big thank you is due my agent, Margaret Hart, of the Humber Literary Agency, whose faith in my work honored me, and whose efforts put me in touch with the University of New Mexico. Lisa Pacheco of the University of New Mexico Press has been my editor and nursed the book through to publication. Copyediting requires patience, perseverance, the ability to persuade, deft handling of someone else's prose, and of course expertise. Karen Brown was the copyeditor for this book and it owes a great deal to her gentle and considered comments, her thoughtfulness, and her commitment to the project.

ONE

Burning Bright

The Turning Point in Human Evolution

Gracie was a chimp at the primate center at Holloman Air Force Base. An old female, but still perky, she had been sent to Holloman from a circus and remembered all the things she had learned and, I guess, enjoyed in that human environment. One of those things was a keen appreciation of tobacco. The chimps lived across a moat from the rest of the base, and Gracie would come to the moat and vocalize at whoever was standing nearby. The intent was apparently to get someone to throw her a lighted cigarette, which she gratefully and gracefully retrieved. She would grasp it like a tube: thumb beneath and two first fingers on top and, smiling, happily puff away. She had other strange habits that she learned in the circus, including making faces and obscene gestures accompanied by vocalizations, but this one really struck me. Here was a chimpanzee smoking a burning cigarette. Certainly she smelled the smoke, felt the heat as she dragged on the lit tobacco, and saw the lighted end. What, I wondered, would make apes—and here I was thinking of them as a prototype of the human ancestor—overcome a fear of things burning and, to the contrary, approach or use it? And if they did approach it, how long would it take before these beings would habitually seek a burning stump to stay near, and then how long till they began to feed the fire until, finally, fire in their lives became something ordinary?

Yet we know that while fire eventually became "ordinary" in the lives of our human ancestors—a vital part of daily activity and survival—it also became one of the most potent of all human symbols and the inspiration for many myths and folklore. How humans acquired fire has been the source of legend through time and across cultures. The association between sun and fire is blatant. There is heat and light. The colors are the same, both burn, and both are essential. Many cultures have stories that tell how the trickster or the champion captured a piece of the sun in order for humanity to have fire. The trickster takes an animal form, often one who seems the weakest or most vulnerable, like Europe's fox, West Africa's spider, or its import to North America as the "trickster" par excellence, rabbit. This being invariably and inevitably succeeds at duping some larger, putatively more powerful animal. Champions, or heroes, on the other hand, are in the mold of Prometheus, who, according to Greek mythology, brought fire to humankind. So great an undertaking requires no less than a superhuman being and often one who defies the greater powers that be. No less often that rebellious hero, once again like Prometheus, is punished throughout eternity for making humanity godlike.

A Choctaw story from Tennessee tells about how "the People" acquired light (Choctaw Tales 2005). In its essence, it recounts how wit and age, here in the person of Grandmother Spider, found a way to trap the light from the sun to relieve the dark world, for in the beginning there was no light. Opossum tried first and ventured to the sun to trap a piece of light. He tucked it under his tail, which burst into flame, so he dropped the piece and returned home empty-handed. Then Buzzard tried, but he kept the piece of sun on his head; he did not succeed in trapping the light because his feathers burned, which is why buzzards have bald, red heads to this day. Finally, old Grandmother Spider journeyed to the place of the sun, marking her way with a trail of spider silk and taking with her a clay bowl. She caught a piece of the sun, put it into her bowl, and managed to bring it back. That is why the People have light, why bowls get baked in fire, and why spiders make webs like the rays of the sun.

The Apache story features another trickster, Fox, the hero who manages to capture fire. He does so at the village of the being Firefly, whose people know fire and can make bonfires. Fox attaches bark to his tail and ignites it. Using a magic cedar tree, he escapes the village, but, as he is running to his burrow, he tires and gives the burning bark to Hawk who, in turn, gives it to Crane who continues to disperse it, the sparks flying hither and yon. Firefly catches up with Fox and pronounces his punishment—while fire is now available to the Apaches, Fox will never be able to make use of it himself (Welker 2005).

The number of tribes in Africa defies counting, and the number of myths per group is a multiple of that. Some tales reflect Muslim influence, while others have a Christian flavor. There are, however, tales that go back to "time out of mind" and that tell of the capture of fire in ways similar to those already described. The Nuer (in Sudan), as reported by the famous anthropologist Evans-Pritchard, had a creation myth in which the original "Eden" was populated by animals who befriended one another and by a human form who never felt hunger since its stomach wandered independently, eating insects. People did not labor, and there was no desire to mate as male and female sex organs lived apart from them and separate from each other. People had no knowledge of fire, nor did they know of or use the spear. Then chaos erupted. In one version, it started with Fox telling Mongoose to throw a club in Elephant's face. As a result of this discord, the sexual organs found their male and female homes, and man and woman began to desire each other. It was Mouse who was instructive in this, teaching man how to make children and woman how to bear them. Fox, the one who brought the dissension, introduced man to the spear and taught him how to use it. That was when the killing began, and the first victims were Cow and Buffalo. In this time of conflict, Dog came and brought fire to the people. In this context, fire has a mixed meaning: while it is a necessity, it also brings pain and destruction; its coals and embers can burn, its flames can destroy a home.

So how this complex, quasireligious relationship with fire began, we can only suppose. There is absolutely no evidence for fire being *controlled* until just over 1 million years ago in Kenya (Bellomo 1994a; 1994b), and just under 1 million years ago in Israel (Goren-Inbar et al. 2004). This newest data confirms that, by this time, a very humanlike species named *Homo erectus* had learned to manage this source of energy. To me, this presupposes an intimate and long-standing connection to, and association with, fire. We also know from Neanderthal burials that the heat and color of fire inspired symbolic parallels to life and death. Here, the dead were rubbed with red ochre, and it is assumed that by smearing the red pigment on their dead, the kinsfolk were symbolically restoring heat—as in the red of sun or fire—and thereby returning life to the deceased. But my question is concerned with beings who could not yet speak, who were just beginning to walk regularly on two legs, who were still using trees for sleeping and feeding while they ventured into drier habitat. What if they had had a relationship with fire? What could they have known about fire? What is the evidence for their association with and use of fire? Which of our ancestors was responsible for conquering fire? How and why did they do it?

I think that the very act of associating with fire was the turning point in human evolution—the "moment" when humanity became an agent as powerful as other forces in nature, such as waves, light, or wind. What happens when dark does not come because firelight intrudes into the night? All organisms have their biological cycles set to light, so how did association with fire affect the daily and annual rhythms, or the reproductive system, or even the "mind"? What would it have meant to the prehumans in terms of that other major property of fire—light? So the most important, and the most difficult, question concerns the light emitted by fire and what that light did to our forebears as it entered the eye and changed from energy to hormones. Brain hormones communicate with the rest of the body, and hormones regulate all sorts of processes: temperature, growth, and cycles, to name a few. What evidence is there for the genetic changes underlying the obvious differences between contemporary apes and ourselves? What are the mechanisms that caused them? What actions performed by our ancestors impelled these changes? And basic to all of these is the question of when the relationship to fire might have begun.

Logically speaking, it would have to have been when there was a "critical mass" of intelligence, behavior, and certain anatomical characteristics—in particular the ability for sustained movement on two legs. The required intelligence includes being conscious of the value of the relationship with fire and to therefore seek it out. The behavioral prerequisites imply actions in the repertoire that reflect intelligence, in particular cultural behaviors—patterns transmitted from one generation to the next. Because behavior does not leave a fossil, inferences about it are drawn from whatever clues can be gathered, including pollen, feces, stone, and the bones themselves. In addition, we have, in a sense, a witness to those eras in the behavior of modern-day chimpanzees—both the smaller bonobos as well as the common chimp—with whom we share ancestors.

Bipedalism is critical. It is a major criterion of "humanness," this ability to *walk* on two legs—not shuffle or run, but walk with a smooth, through gait. In the time of Buffon, one of the greatest eighteenth-century naturalists, anthropoid primates were called—indifferently—quadrupeds or quadrumanous, recognizing that the front "feet" were indeed hands, manipulators as well as locomotors. Bipedality would have meant that the hands were free to carry a desired object from one site to another. Monkeys do *sit* erect, freeing their forelimbs for various tasks involved in getting food and for social behaviors like grooming. Modern apes habitually locomote on minimally three limbs, they will stand and run bipedally, but they cannot *walk* bipedally—they do not have the anatomical requisites: the Achilles tendon and certain vertebrae (Filler 2007). Find a bipedal hominiform fossil (one that looks human), and you find a creature that represents a stage clearly on the way to becoming human. Good evidence of this now exists at 6 mya (or 6 million years ago), and some argue that various *different* forms of bipedalism existed as far back as 15 mya. The body plan in the lower back permitting bipedalism was already present around 20 mya (Filler 2007), although *efficient* human walking probably did not occur until around 3.5 mya (Sellers et al. 2005; Wang et al. 2004). (A note on dating: mya—megaannum—and kya—kiloannum—are the International Standards convention ISO 31–1, standing for millions of years and thousands of years respectively.) Over the millennia, bipedalism was progressively favored by natural selection (Sellers et al. 2003; Sellers et al. 2005), as witness the development of a longer lumbar curvature and changes in vertebrae to compensate for the front load in pregnancy (Whitcome et al. 2007). There is now increasingly clear genetic evidence as to which genes, which genetic mechanisms (Calarco et al. 2007), and which rates of mutation (Hahn et al. 2007) spurred the evolution of genetic differences that brought about the physiological, anatomical, and hormonal divergence between apes and humans.

There are three major periods of time in which intelligence, anatomy, and behavior could have reached the "critical mass" suggested earlier. The first, known as the divergence, is the span of time during which each of two branches of apelike beings changed in different ways: apes to apes in one pattern, apes to ancestors of humans in another. This is thought to have happened somewhere between as early as 10 million years ago to as late as 5 million years ago. (A note on terminology is important here. For the purposes of this book, I will be referring, from here onward in the text, to the human "Ancestor" of this first period with a capital *A*. This informal designation, not part of scientific terminology, will encompass the beings that were ancestral to humans only at the time of the divergence from apes, that is, somewhere in the broad expanse of time between 5 and 10 million years ago.) The second period in which this relationship might have begun is somewhere between 4 to 3.5 mya. In this million-and-a-half years, an "explosion" of types within the ancestral group adapted to different conditions. This is the most logical period in which to place the start of the association with fire because there is sufficient fossil evidence about anatomy to infer the sort of intelligence and behavior needed to meet the criteria given above: bipedality, consciousness, culture. It is even more tempting, however, to place the events surrounding the relationship with fire much closer to our own time, in the third and best-known period of transition—around 2.6 to 1 million years ago. This is the time of ancestral relatives known as late australopiths (Wood and Richmond 2000) and the emergence of *Homo*, the genus to which we belong, and we know a lot about these people.

At 2.6 to 1 mya, there are enough fossils to present a picture of group life and the range in forms—male, female, old, young—and to make clearer interpretations about diet and behavior. In effect, we can determine just about their entire way of life. The oldest stone tools date from 2.6 mya, and many other events occurred at this time as well—genetic mutations in jaw muscles, which indicate changes in diet, posture, and facial structure (Stedman et al. 2004), and in cognitive abilities—which makes it appealing to place the control of fire within this interval. It seems to me, however, that this period marks *achievement*: the basic "work" of becoming human has already been accomplished. Surely it will have taken a long, long time to get that way, and a lack of evidence from the earliest period does not indicate that nothing was going on! Stone tools leave a trace, but bone tools would hardly be distinguishable from some animal's dinner, and wooden ones rot to dust. Indeed, the earliest tool of wood that has been found more or less intact is a spear. It was found in Germany, and while it is incredible to think that a shaped and used piece of wood could last from 400 kya (400,000 years ago), the spear has been dated to that time (Thieme 1997). The find is impressive, but deceptively so.

Watching nonhuman primates, especially the apes, gives testimony to the abilities that must have existed before language developed (see figure 1). We know that species like monkeys, which are even more primitive than apes, use wood as hurling objects, and apes are known to fashion tools from wood (Goodall 1986; Boesch and Boesch 1990). They *use* twigs, sticks, and even bark in their "expedient technology" as tools with which to dig tubers from the soil to a depth of nearly 25 cm (10 in) (Hernandez-Aguilar et al. 2007). For more mobile food, these apes *craft* sticks and twigs to capture social insects, like ants or termites (McGrew et al. 2005). They have an inventory of twenty-six different tools, and they choose their spears when they go to hunt galagos (Pruetz and Bertolani 2007) or squirrels (Huffman and Kalunde 1993) in the forest, poking into the nest cavity of the tiny primate. It stands to reason, therefore, that wood would have been used as some kind of tool long before the first evidence of it. Being organic, wood is subject to rapid decay, so it is unlikely that archaeologists would find such tools—even if the Ancestor made them—in the earliest time horizons. By the same reasoning, fire too would have become part of everyday existence long before any trace of that phenomenon ever left evidence of its use.

The profound similarities between apes and humans and the equally stunning differences between us indicate that something major occurred in our line at the divergence accompanying the ability to walk on two legs. The ability to climb was not yet forsaken, however; evidence of long forelimbs suggests that a facility in the trees accompanied the first forays in bipedalism. It is for these genetic and anatomical reasons that I am setting the scenario of fire association and use over the long period during which apes and humans diverged, but after the first appearance of bipedalism, that is from 6 to 4 million years ago. This recognizes that the Ancestor in this early time frame had certain qualities: freedom of

FIGURE 1. Adult male bonobo attempts to crack open a nut with a stone tool. Courtesy of Pamela Heidi Douglas and the LuiKotale Bonobo Project, Democratic Republic of the Congo.

ONE

the forelimbs, erect posture, at *least* the equivalent of ape intelligence to experiment and assimilate experience. There appear to be stages in the millions of years of relationship to fire, beginning with the willingness to approach, sit near, tend and use, and ultimately to manufacture fire. It is this latter achievement—the domestication of fire—that is a hallmark of our species, *Homo sapiens*. But "humanness," or hominization, is a process. It began slowly but each step caused the next one to accelerate over the vast tracts of geological time. Change triggered further change at ever increasing speed, and once past a critical threshold, the iterations accelerated ever faster. I see the advent of fire in humanity's life as a consequence of who they were and a precipitator of who they were to become. Fire was not a by-product, nor simply another influence: it was *a*, or maybe *the*, major contributor to the manifold changes that made us human.

What could have made the Ancestor want to approach fire? Because the Ancestor did not rely on eating or hunting big-game animals, other foods, especially fat- and protein-rich insects, were probably a significant part of the diet. Insects continue to be amongst the favorite foods of nonhuman primates. They eat a huge variety of these, although the most voracious insect eaters are New World monkeys and prosimians—the earliest forms to evolve in the primate group, suggesting a long and well-established dietary habit. This behavior is also well documented in Old World monkeys. They too eat a considerable variety of insects and are quite adept at plucking them from the air. They are equally capable of discriminating between edible and foul. Apes tend to be particularly interested in termites, and among chimpanzees it is mostly females who pursue the termite fishing pastime with twigs they shape to poke easily into the entrances of the nests.

Certain fire-loving insect species congregate near embers, attracting other creatures to their harvest. The desirability of these insects could well have tempted the Ancestor—like modern-day chimps—to prowl the edge of a smoldering fire to pluck out some succulent bug. Modern apes will, under special conditions, approach a burning fire, or even imitate how humans use it. Finding this behavioral accommodation in contemporary apes suggests that familiarity with fire, and use of fire, were behaviors beginning to take place as humans and apes diverged. Fear of the heat and flames of an active fire in animals is quite common, although, importantly, it is not universal. A cooling fire, reduced to embers and coals, on the other hand, attracts birds and other animals looking for charcoal as medicine (Struhsaker et al. 1997) or prey that got trapped and roasted. The inclination of all primates to gorge on insects whenever possible, whether flying ants or termites, grasshoppers, and other bugs and beetles, makes it reasonable to assume that the Ancestor was similarly attracted, as are a variety of peoples around the planet today. There are insects, known as pyrophilous or fire-loving insects, which abound in and around a cooling fire. An ape-like creature with a love of nourishing insects would certainly know their availability, and where and when to find them. The absence of fear enabled them to approach and tolerate burning fires where they slept and ate.

It is the premise of this book that proximity to fire stretched out the period of light and irreparably altered hormonal cycles that are dependent on light and darkness. Understanding the impact of firelight on the brain means delving into how light gets into the brain, how the eye works, and its anatomical and hormonal connections to structures deep inside the brain. This means understanding light itself, at least as far as campfire flame is concerned—a daunting inquiry at best. Is the spectrum of light in a campfire sufficient to stimulate sensors in the eye and brain? Over eons, would the impact of firelight at night affect the brain to the point where cycles dependent on the day-night contrast would be altered? Would the physiology of reproduction change as a consequence of increased exposure to light and its hormonal consequences, and with it social relationships and roles?

Certainly, changing the light-dark cycles had major implications for social structure in prehumans who, like other primates, were social animals. Primates tend to develop organization and structures in and around breeding rights and parental care. Social animals live by rules and tend to have developed cognitive processes: "mind" and "thought," which depend on memory. In turn, these social and cognitive characteristics impact settlement patterns, food-sourcing patterns, division of labor, and on and on. Changing light-dark cycles has major implications for these, as melatonin, a hormone involved in the regulation of biological rhythms, seems to interfere with memory formation (at least in zebrafish) (Rawashdeh et al. 2007). Extrapolating from their findings, I wonder if firelight, by inhibiting melatonin, enhanced memory formation and associative patterns in the Ancestor.

Perhaps even language acquisition was enhanced by the extended day afforded by the fire. Not that fire inspired the ability to speak, but it may have provided a place and the conditions that created and encouraged communication in connected tones more specific than gestures or single sounds. Imagine the evening: the sun having set, the heat of the day is dispelled in the breeze that rustles the branches of trees. The sounds of crickets, bats, and nocturnal animals punctuate the night. The sounds beyond the crackling of the wood in the fire create a comforting place of this area, bathed as it is in the firelight. Prehumans sit there, tired, playing with children, grooming each other, removing dirt, bits of twigs, leaves, ticks, and other biting insects. Certainly the complex communication systems observed in our primate cousins permit us to imagine an exchange of objective information ("This tree is in fruit"), as well as subjective information ("I am thrilled about the ripe fruit"). Increased reliance on communicating *information* itself fostered development of structures already present. Over ages, the need to transmit information also promoted genetic selection of anatomical structures that enhanced both nuance and precision. As human consciousness developed, the *meaning* of fire and firelight expanded (Rilling et al. 2008), and all sorts of power and volition were attributed to it. As social animals that could speak, modern humans invested events and phenomena around them with symbolism. Recognition of fire's power led to tales of magic and awe and divided the world into light and dark and *forces* of light and dark. The firelight dispelled whatever it was that lurked beyond the shadows.

The impact of fire on human evolution is complex, and there is much to understand, connect, and process. Every fact leads to a new question, every fact is debatable, and every fact is an exciting contribution to a picture of a creature moving irrevocably along a trajectory over which it gains greater control. Because I have studied monkeys for over thirty years, I have developed an understanding—part intuition, part analysis—of what inspires their actions and, more significantly, their reactions. I know what monkeys are capable of, be it their motor skills or their intellectual limitations. And yet the facts of monkey life are often obscured by the nature of their bigger relatives, the apes.

The genetic similarity of apes to us helps promote the idea that ape behavior is basic to human behavior. But monkey behavior is basic to ape *and* human behavior. What monkeys can do warrants attention in terms of reconstructing the mentality of the ancient hominoids and their descendants, the hominids, closer and closer to the hominins that include us and our direct ancestors. A word on new vocabulary is needed here. There has been a profusion and consequent confusion of names referring to humans and their relatives (see, for example, Underdown 2006). These terms all begin with "homin." For the sake of clarity, and in keeping with modern scientific parlance as per the following citation, I am using "hominin" to include chimps, gorillas, humans, and their direct ancestors back to 6 mya, referred to as "early hominins" by scientists following the work of Wood and Richmond (2000) and Wood and Lonergan (2008).

> [F]or the family Hominidae has become more inclusive, and now refers to the common ancestor of the living African apes (i.e., *Homo*, *Pan*, and *Gorilla*) and all of its descendants. The appropriate vernacular term for a member of the human clade is now "hominin." (Wood and Richmond 2000, 20)

At time horizons in the millions of years ago, the behavior of our ancestors can only be inferred. Most of the reconstruction depends on behavioral observations of contemporary apes and monkeys. (There is simply no other model to go by.) In the heyday of the apes, the time called the Miocene, nearly 25 to 5 million years ago, there were some tens of different kinds of apes roaming from Asia through Eurasia to Africa; now there are only gorillas, common and bonobo chimpanzees in Africa, and orangutans in Asia. There is no other living form that can provide a mental image to project back in time.

Primates—including humans—differ from other mammals in the nature of their brain and the capabilities that structure permits. Darwin already noted that humans are less reliant on teeth and nails than are other mammals (Darwin 1871). Evolution in the human line is characterized, instead, by increasing reliance on the brain, its product "mind," and the further development of mind, the construct "culture." Innovations in the brain in turn affect anatomical structure and function, which feed back to the brain. Mind and culture became evolutionary pressures selecting for more mind and brain. Primates, as a group, are acknowledged as being flexible organisms, able to adapt to local demands—environmental and social—not only through genetic evolution but also, importantly, through cultural processes devised in group living. This behavioral system is faster than physical evolution, although it requires flexible systems of communication. Nonhuman primates use sounds and gestures to ensure the accurate transmission of social behaviors (Dunbar 1996). A big brain with lots of storage space, aided by a system of recall and quick association of distinct memories, has been the hallmark of the primate-to-human evolutionary pathway.

The premium placed on mind begins early in primate history and accelerates with each new form of primate. That prosimians, those smaller primates like lemurs and bush babies, gave rise to subsequent types of primates is now accepted fact. The evolutionary events that conferred on the South Asian tarsierlike prosimian a flat face with overlapping fields of vision also propelled relatively rapid development of the brain. Mind resides in the brain and depends on the cortex to store images, thoughts, ideas, and feelings, all of which are life experiences. That is, they are not "hard-wired" in the sense that the term *instinct* was meant to indicate. What is required in this storage facility is a system that retrieves memories and joins one to another rapidly, thereby allowing purposeful action. Present, past, and future have meaning to humans, and, in a nascent form, to chimps as well; there are, after all, not only differences in the action of genes, but in the development of cerebral structures those genes control (Oldham et al. 2006).

What happens in the brain is so complex that there are many different scientific disciplines to cover it all. The infinitely tiny aspect of this question that particularly concerns me is the interaction between the conquest of fire as an adaptive tool and its effects on the development of mind. This relationship of mind to the acquisition of fire is basic to understanding the process of becoming human. The scenario accounting for this interaction relies in part on the inherent adventurousness of young primates. The more daring amongst our prehuman forebears would venture toward the cinders that still had flares of flame here and there—monkeys and ape children are like that—picking up burning twigs and flinging coals. Given the dynamics of cultural transmission in these primates, such behavior would make subsequent generations even more enterprising. I have frequently observed that monkeys readily overcome unease through play. They hurl themselves into space from high, often dead, trees to plunge into deep lakes and then swim to shore; they leap across huge expanses, legs flailing in a superhero sort of way, to reach a branch or rock overhang an impossible distance away. Apes are no less inclined to wild play. Adults lob branches in the "rain dance"; young apes throw branches from trees, swing on vines, race down hillsides and back up again. It is the *role* of the young to introduce innovation (Burton 1977). What could be more innovative than combining previously practiced behaviors with a fiery stalk or a burning twig? Youngsters grow up observing adults and each other. In turn, as adults, they use the behaviors they witnessed and memorized. These memories extend the repertoire, and in turn become the basis for new behaviors. This may not be progress, but it is change. And in a system, a small change in one place can force accommodation to that change in other parts of the system.

The most solid evidence for cohabitation with fire goes back to the site of Koobi Fora, in Kenya, dated at 1.6 mya (see chapter 6). The next closest is a hearth found at Swartkrans, in South Africa, and dated to 1.5 mya—virtually modern history! But whether the fire was made or brought to the hearth is unknown. There is no sure way of knowing how manufacture began or when. Stone tool manufacture, however, has a million-year history by the time of these sites. Perhaps the errant spark as stone struck stone led to repeating this technique over dried moss or leaves. Certainly the final *stage*, the manufacture of fire, appears to be recent—perhaps accompanying our type of human, *Homo sapiens*, after 250,000 years ago. This is known from materials the archaeologist has found, the physicist has dated, and the electron microscope has analyzed, so that it is certain the artifact is old and of a certain date and made by someone's hand in a manufactured fire.

But *use* of fire is a great deal older. Some scholars consider it an *emergent* behavior because nonhuman primates do not display behavior that could be considered ancestral to fire use (Ronen 1998; Rolland 2004). We know from contemporary accounts that people used to care for fire, carrying embers from hearth to hearth, reigniting torches by setting fire to grass on long journeys, igniting tree stumps to keep the fire going, borrowing embers from neighbors, and even giving embers as a wedding gift—all without rekindling the source. Like any major technological advance (i.e., the cell phone or the microwave), as soon as it is acquired it is virtually impossible to give up. Once approaching fire was the norm, and associating with fire became increasingly habitual, the care of fire, the feeding of it, and the transporting of it would have become as essential to a family group as recharging batteries for the cell phone. Consider the benefits of fire: in a study of contemporary foragers and nomads, Worthman and Melby (2002) found that besides light and heat, the smoke fumigates the home, keeping insects, rodents, and snakes away from thatched roofs. It offers protection from predators, particularly where homes do not have solid walls, such as amongst these peoples. In so doing, it promotes a sense of security. An ancillary benefit might be, these scholars suggest, the vigilance, comfort, and stimulation provided by caring for the fire, sitting by it, and watching the flames (2002, 75–76).

So what is needed is evidence of *hearths*—of fires strategically located wherever the Ancestor slept. The kind of evidence archaeologists look for includes charcoal, sediment that has been baked and is therefore hardened and discolored, bones or stones that have been changed by heat, and—best of all—some kind of structural feature like a hearth, recognizable by a concentration of ash and charcoal in a limited area, and sometimes bordered by rocks. A recent find in Israel establishes just such concentrations in specific places; possibly these were actual hearths. In addition, six different kinds of wood were burned at the site; two of these are from edible plants: wild grape and the olive. What is more, seeds were found of plants important in the diet like wild barley (Goren-Inbar et al. 2004). There are tempting hints of hearths or at least campfires going back toward 2 million years ago. The difference is important. A hearth implies a home site; a campfire is temporary. Archaeological evidence becomes startlingly difficult to find at these early dates. But we do know there were bipedal hominins at 6 mya. What might their relationship to fire have been? Native peoples before the cultural present (1965) carried firebrands from sleeping place to sleeping place. There they used the firestick to ignite a campfire and perhaps stay a few days. They may have constructed a temporary lean-to, or may not have bothered, depending on circumstances. They left and started the process once again. Evidence of these temporary encampments blows, drifts, gets washed away. Could the Ancestors have used fire but left no trace (at least a trace we cannot find)?

Owning fire would have meant the ability to make more fire—to light up the evening sky by setting fire to nearby grasses or trees, to build a bonfire to keep away stalking predators, to define "inside" from "outside" the circle of light. Caves would no longer have been unapproachable and migration to colder zones would have become possible (Rolland 2004; Gowlett 2006). Ultimately, after millions of years of association with fire, access to foods that were inedible became possible. Monkeys dig into the ground to 10 cm or so to find corms, bulbs, and tubers; apes go further (Hernandez-Aguilar et al. 2007). These fleshy storage organs of plants provide good nutrition. Some, however, are so large or so hard that they cannot be eaten, or contain compounds that make the consumer ill. Placing them in the embers of a fire changes that. Now they are softened, the pulp accessible, and the toxins removed. Then too, the cellulose of the plant itself would be changed with heat, permitting the contents of the cells to be assimilated. The achievement of roasting tubers is most likely associated with *Homo erectus* at times a good deal closer to the present (Wrangham 2001).

But what about meat? Meat has not been included as "newly" acquired in the dietary repertoire. Many Old World (Africa, Asia) monkeys—macaques, vervets, and baboons, to name but three—do eat meat. And so do the even older, (in a paleontological sense) New World (Central and South American) monkeys (capuchins and howler monkeys). Eating meat has been around a long time. The meat eaten is from small vertebrates, like lizards. That term—*meat*—is reserved for taking flesh. The issue of "what's in a name?" here creates a strong bias in the study of becoming human. It is not nutrition from protein or fat that becomes the issue, but rather the package these nutrients are found in. I wonder, though, if that was an issue to the Ancestor. Insects are not categorized as meat, although they *are* nutritively similar. Indeed, the winged form of several species of termites were found to contain between 613 and 761 kcal of fat, while three species of African caterpillars averaged just under 500 kcal of fat. These insects were equally stuffed with vitamins and minerals, and although low in some amino acids, they are equally high in others. Taking off the outside "shell"—that is, the chitin exoskeleton—increases the quality of protein so that it is quite comparable to "meat" from vertebrates (DeFoliart 1992). Retrieving insects from the embers of natural fires might have achieved this for them.

Meat in one form or another, then, has long been utilized by primates. What the hominins would have added to the dinner was the *cooking* of it. How that began is, of course, unknowable, although Charles Lamb's essay "On Eating Roast Pig" has long been cited as an amusing but rational supposition. But his method requires that people lived in huts, had rope of some sort, and had already domesticated the pig that ended up roasted when the hut caught fire. Still, the idea of an accident eventually resulting in something useful underlies the story, and that may have happened millennia before there were huts or rope. Perhaps someone did drop a piece of meat into the fire; perhaps someone toasted a piece of flesh on purpose because insects tasted so much better that way. In any case, cooking occurred at some ancient time, but evidence is more difficult to find. Recently some scholars suggested that meat eating had to have *begun* at 1.9 mya because there is increasing evidence of big-game hunting with *Homo erectus*, and mammalian meat would have been too fibrous and tough to be eaten with the teeth and jaws that these near-humans had (Wrangham and Conklin-Brittain 2003). In addition, fossil evidence from *Homo ergaster* indicates that tapeworms derived from animals had infested this fossil hominin by 1.7 million years ago (Hoberg et al. 2001), which suggests that the cooking was either not done long enough to kill the parasitic larvae or the fire was not hot enough. It also means that there was eating of mammals. While big-game meat eating may have required the skills, organization, and cognitive abilities of *Homo*, the process of utilizing fire to alter food, I suggest, would have begun as early as the Ancestor was using fire. Indeed, if they did begin with gathering the fire-loving, pyrophilous insects, the connection would have been made there, at approximately 6 million years ago.

The foregoing discussion underscores just how many different ways life changed for the Ancestor who learned to use and control fire—physically, mentally, environmentally, and socially. There is little dispute about that. In this book, though, I am proposing a much earlier date for the first ancestral *contact* with fire than has generally been entertained, a proposal that will surely be controversial given that archaeological evidence for the *control* of fire suggests a date no earlier than approximately 1.6 million years ago (at Koobi Fora, in Kenya). I will argue, however, that such a date represents a much later point toward finalization of both hominization *and* control of fire. I think that the very act of associating with fire occurred much earlier and was in fact the turning point in human evolution—the "moment" when humanity became a powerful agent of its own continued creation. Thus, in thinking about the whole process of hominization and the role fire played in it, I have begun to see these earlier hominins, not the later members of the genus *Homo*, as being responsible for the first steps in the domestication of fire.

The significance of these first steps lies in the subsequent influence they had on the course of human evolution itself. I suggest that the more that these hominins of approximately 6 million years ago *did* something to the environment by way of intervening with natural processes—in this case, artificially increasing the daily hours of exposure to light—the more they dampened the effect of the environment on themselves. In consequence, they took over the direction of their own evolution on many important fronts, the most significant of which may be the development of mind. The acquisition of fire is precisely the kind of intervention that could have initiated and perpetuated this chain of events. The chapters that follow detail the interplay of who, where, what, when, how, and a little of why, as they focus on the question of how fire influenced human evolution.

The Anatomy of Fire

What Made the Ancestor Approach?

Fire can be dangerous, lethal, and horrifying in its potential for devastation. So why do any creatures associate with it at all? Why and how did the Ancestor begin and gradually deepen a relationship with fire? There are two major questions here: (1) the ecological role of fire in the African savanna, and (2) what we know of contemporary, nonhuman primates—their diet, their cognitive and social capacities, and their likely motivations to approach fire, as well as observations of their experiences with fire. These questions lead to a model of the steps the Ancestor might have traveled in first identifying the beneficial uses of fire: proximity with it, learning about its properties, containing and nurturing it, and ultimately learning to manufacture it.

The African Savanna

The larger, environmental context within which all of this experience and learning would have come about, and in which the Ancestor was evolving, was the African savanna in the Miocene. The Miocene as a geological period began about 23 million years ago and has long been considered the time of grasslands. Recent evidence from phytoliths—little stonelike fragments in plants—indicates that grasses originated and had already diversified millions of years before primates were even born (Piperno and Sues 2005). Knowing that these grass environments existed much, much earlier than previously thought is new and significant in extending the habitat for the earliest hominins and the fauna they lived with.

The African savanna, historically and still today, experiences such widespread burning as part of its normal cycle of devastation and regeneration that Africa has long been called the "fire continent" (Govender 2003; Keeley et al. 2005). How widespread is the savanna ecosystem in Africa, and what role does fire play in its maintenance? Savannas are now spread over almost 50 percent of the landmass of Africa—some say 65 percent (although much is giving way to desert). By definition, these are zones where grasses and trees are found together. Savannas can range from mostly grasses to mostly parkland, depending on the openness, and are complemented by gallery forests that run along streams. These permit animals to retain their forest adaptations while exploiting the neighboring savanna. Annual rainfall in savannas can be quite low—as little as 15.24 cm (6 in) has been recorded. The amount usually ranges from 51 to 127 cm (20–50 in) per year, although as much as 25.4 cm (10 in) has been noted in one rainfall. The alternation of wet and dry periods is essential to the maintenance of savanna. Too much rain, and the area becomes forested; too little, and it becomes desert. Rainfall is concentrated in 6 to 8 months of the year, after which a long dry period follows, bringing fires with it. The dry season provides the tinder. The grassy understory dries and provides the extremely flammable fodder—the plant material that has dried throughout those months (FAO 2001). Current research indicates that 2 to 3 metric tons of tinder per hectare is necessary for fire to get started (Danthu et al. 2003). Lightning is the primary ignition source maintaining the savanna system. Since the mid 1990s alone, reports of lightning striking and killing a giraffe (southern Africa), two tea-station workers, and several prisoners, among others, have been reported (Govender 2003). Currently, according to NASA's new lightning detectors, central Africa is the area receiving the most strikes, although southern Africa is receiving more than its share.

Although our image of a grassland fire is one of a virtual holocaust of flame, smoke, and devastation, under natural conditions—that is, where invasive plants have not yet gained a foothold—the savanna community is hardly devastated. Observations made in southern Africa have shown that the temperature of fire grows hotter where there are "alien" plants and that animals are more often burned under these kinds of conditions. Seeds, too, are devastated in the increased heat (Farm 2006). Some seeds, however, profit from the heat to split open, some are suspended above the grass and are relatively untouched, while others are buried within the soil where the high temperatures do not reach them (Danthu et al. 2003). Furthermore, in a "natural" burn, fire burns near the ground in a discontinuous front, resulting in a mosaic of burned and unburned areas that provide shelter and food. While the dry matter of plants—leaves and stalks—is consumed by fire, the roots, tucked under the ground, are not. These contain considerable amounts of starch as well as moisture and are the source of new growth. Savanna trees tend to be fire resistant, this benefit being conferred on them by their corklike bark or a trunk that has a good deal of resin in it, the moisture protecting the tree from harm. Some trees use chemical protection: the bark of the mopane tree, common in southern Africa, contains crystals of calcium oxalate, a natural fire retardant. At high temperatures (around 370°C), the crystals decompose and give off carbon dioxide (Naskrecki 2005).

Fire burns the litter layer of debris on the ground, killing a variety of insects and their larvae. Although insects are the most affected by fire, they return nutrients to the community, beginning with plants that profit from the additional nitrogen available from burned insect exoskeletons and ending with the minerals and other nutrients that have been returned to the earth in the drifting ash. The burned area may look like the surface of the moon, scorched and blackened, but it provides a veritable feast for a variety of animals. While the ash is still hot, birds and mammals—especially nonhuman primates—come to seek newly available food resources like seeds and nuts, opened by the heat. Insects like grasshoppers, stick insects, and beetles are plentiful, and small vertebrates like mice and lizards either become visible due to the loss of cover or are already dead. A variety of insects, the fire-loving pyrophilous insects, are attracted to burn areas because of the nitrogen, and can arrive at a fire site within 24 hours, oftentimes in great numbers (Whitehouse 2000). Among these fire-loving insects are those that go out of their way to consume dead or dying wood (Whitehouse 2000), the saproxylic insects. Of these, perhaps the most important are the termites, whose huge mounds are architectural marvels on the savanna. In some parts of eastern Africa, these mounds provide soil for tree growth; elsewhere, they are virtual banks of seeds, nuts, and other food sources that the insects have hoarded (House and Hall 2005).

When the rains come, there is new growth. The storage organs—the roots of shrubs and grasses—are a source of the energy that the plant needs in order to grow once new rains have come and drawn burned nutrients down into the soil. An astounding 2.54 cm (1 in) of growth in 24 hours has been recorded for some of the larger grasses. Shrubs also have deep roots and can live from the nutrients stored in them until the appropriate time to begin growing. Hence, fire encourages vegetative diversity in general (USGS 2006a) and the return of insects in particular (Whitehouse 2000). Savannas evolved with and because of fire and over millions of years promoted a fauna of grazers and browsers well adapted to this habitat.

Why Approach Fire?

Clearly, the devastation and havoc wrought by fire's fury inspires awe and fear. It makes sense that an animal should develop fears in reaction to dangerous situations and that these should somehow become "coded" into its patterns of response. Under normal conditions, animals are at least aware of fire and can get out of harm's way. Frogs, for example, are sensitive to the sound of fire and flee before it (Grafe et al. 2002). Small mammals are frightened of fire, as they cannot easily outrun one. They therefore suffer from its effects in ways that large mammals do not. Although most find refuge and survive in underground dens, caves, and rock crevices, these burrows can be depleted of oxygen and their denizens killed as the fire gains momentum.

For large mammals in forested areas, where fires tend to be infrequent, fire can be lethal. Large mammals that live in *grasslands*, however, are typically not afraid of fire (Whitehouse 2000) due to the fact that they receive cues from scent and sound, giving them sufficient advance warning to avoid it. This adaptation is thought to be because brush fires are benevolent with respect to them, affording new growth as well as maintaining diversity in plant species. In addition, fire reduces pests that plague mammals, ticks and other ectoparasites in particular. Field ecologists studying white-tailed deer in North America observed that they would feed within 20 m (65 ft) of approaching fire without showing any alarm and were never seen to run away from it (USGS 2006b). This response to fire, however, is labile; where there has been no burn for as little as fifty years, the response of these large mammals to fire is once again one of fear.

Primates, too, vary in their responses to fire. Baboons take to natural gorges where the moisture in the trees affords protection. They have been observed to feed calmly in the face of approaching fire and then, as the fire neared, to double back behind it to feast on the acacia seeds and other pods that had opened with the heat (Gaynor 2006). One type of Philippine tarsius, a small primate with remarkably large, nocturnal-adapted eyes, is called "carbonarius," apparently from its habit of picking burning embers from the fire (Oakley 1961), although I wonder if it was actually insects that it was seeking.

In yet another example of grassland fauna recognizing the usefulness of fire, we can observe that smoke serves to signal birds of prey, which fly toward the fire and either circle while waiting for the flames to diminish or fly before it to capture the fleeing animals. After the burn, the area is devoid of cover, and the birds can simply pick off their dinner. David— the—David Livingstone, the explorer, is attributed with observing hawks and other birds in Botswana grabbing insects and small mammals as they ran in front of a bush fire. Rooks and crows, according to one authority, fly in front of a fire with wings outstretched, apparently trying to smoke the bugs out, and cattle will gather at a fire made of dung to keep mosquitoes and other biting things away from their skin (Clark and Harris 1985). Predatory mammals, knowing that fire flushes game before it, approach the fire front. For mammals that eat insects, a burned-over area provides a bonanza since these locations have been found to contain greater numbers of arthropods than unburned areas (USGS 2006b;). Birds, among others, visit burned areas to gorge on the abundant availability of insects. In the southeastern United States, one such bird, the bobwhite quail, has been nicknamed the "fire bird" because it will rush to the edge of a burn before the fire has even stopped smoking to partake of a feast of dead insects and seeds (USGS 2006b). It is a curious fact that grasshoppers change color to match the burned land after a fire. Fire melanism affords them camouflage from predators.

Thus one important benefit of grassland fires to large mammals and nonhuman primates is as a source of food. Birds, mammals, and nonhuman primates arrive at a burned-over area seeking the newly available food resources such as seeds and nuts, opened by the heat, as well as the insects and small vertebrates like mice and lizards that either become visible once the groundcover is burned away or that have already been killed by the fire.

For the Ancestors to have been willing to recognize and to associate—and that is the key behavior—with fire as a valuable food resource, they would have needed to share three behaviors that we already know to be documented in monkeys as well as apes. First, they would have had to be able to *remember* the location where food had been found at least once—an ability shared with a number of mammals. Second, they would have had to eat earth and charcoal as digestive aids (Burton et al. 1999; Engel 2002). Geophagy, the eating of soil or charcoal, is well documented in a variety of mammals and birds (Struhsaker et al. 1997; Bolton et al. 1998; Klein et al. 2008). Current thinking is that this behavior permits the eating of foods not ordinarily preferred that may contain substances ordinarily toxic or that interfere with digestion (Burton et al. 1999). Eating soil or charcoal decreases the amount or intensity of these substances and thereby makes available plants and insects that would otherwise be ignored. This would have been important to the Ancestor because, by the time of the Late Miocene, a wet warming trend, with an annual dry season lasting some 3 months, began to provide more food from plants like grasses, their seeds, stalks, and rootlets. The evolving hominin, however, had to adapt as these plant substances interacted with their physiology. Since plants contain a variety of secondary, variously toxic, compounds that serve to protect them from insects and other mammalian predators (Stahl 1984), primates (and other animals) need to eat charcoal (Struhsaker 1997) and, more typically, soil, which either neutralize or absorb substances that would otherwise make them ill (Burton et al. 1999).

Third, the Ancestor would have had to eat (even relish) *insects*. I have seen monkeys in Africa rush to get at swarming insects in order to bite the succulent sweetness from the abdomen. Macaques in Asia pluck insects in flight uncannily quickly and devour them, looking for more. I have even seen a Tibetan macaque grab a spider and put it in its cheek pouch, where, I suppose, the poor wretch drowned in the saliva. We know that nonhuman primates are as a group omnivorous and have the ability to ingest nutrients from a wide variety of substances. The evidence for this is both observational (from contemporary nonhuman primates) and anatomical (in that all species except prosimians have four kinds of teeth, the shape of each allowing access to various food types). The pressure to become omnivorous increased as food sources became unpredictable (through periods of drought, for example), and the majority of nonhuman primates became able to exploit diverse diets. Anatomical comparison between the teeth of early hominins and their earlier cousins provides evidence of this omnivorous trend (Verhaegen and Puech 2000). The molar teeth of the early hominins had thicker enamel, the jawbone was more robust, the molars were lower, and the incisors wider. Pitting on the incisors as well as striations on the molars indicate that abrasive vegetation was being used. The gritty material would have been found closer to the ground— that is, the hominins were not feeding from the trees as their more apelike ancestors had done. These microwear patterns on hominin teeth clearly indicate that the diet was coming from open areas. The early Ancestor was eating an eclectic diet.

Other dietary pressures came to bear as the genetic ability to synthesize vitamins C and B12 was gradually lost across nonhuman primates (occurring as the genes controlling the manufacture of these vitamins were appropriated for other functions). Vitamin C comes primarily from fruits and vegetables and is far more readily available to nonhuman primates than B12. Vitamin B12 (cobalamin) is required to prevent cobalt deficiency and is essential in the production of blood cells. The primary source of B12 is meat or, in the case of nonhuman primates, insects (Wakayama et al. 1984) and is not readily available from plant sources unless these have been contaminated by fecal matter. Originally, primates relied on B12 manufactured by bacteria in the gut, as vegans still do, but as the genes manufacturing this vitamin were appropriated for other purposes, primates turned to insects to get this required vitamin. In one study, five termite species produced .455–3.21 mcg/mcg—a substantial amount of this essential vitamin (Wakayama et al. 1984). B12 has been found to enhance shifts in human daily rhythms, mediated by the hormone melatonin (Honma et al.1992; Hashimoto et al. 1996), and actually enhances the action of melatonin (Ikeda et al. 1998).

With the loss of the ability to self-synthesize these vitamins, it became increasingly vital for hominins to increase the amount of meat in their diet—primarily by eating insects. Bill McGrew and colleagues (2005) have written extensively on the importance of insects in the diet of apes, and I extend that importance to Old World monkeys also. As suggested above, insects rival meat—that is, muscle—in the nutrients they contain. They are high in fat and protein, with a dash of minerals as well. Insects can be tasty, nutritious, and filling. I experienced this myself when my daughters once insisted that we deal with the grasshoppers attacking our vegetable garden by eating them. We followed a recipe they had seen on a children's television program, which involved first boiling the creatures, then removing the legs, then quick-frying them in butter and garlic. The grasshoppers were incredibly rich—it only took a few to feel full. Extensive research has been done on the nutritional value of a wide variety of insects. The protein, fats, minerals, vitamins and even carbohydrates (in the carapace) make them a valuable food source (DeFoliart 1992).

Humans and their nonhuman relatives have always utilized insects to varying degrees. There is even firm archaeological evidence that *Australopithecus robustus* used tools to extract termites (Backwell and d'Errico 2001), much in the manner of contemporary chimps. More recently, there is data from the shells in human feces found by the Great Salt Lake indicating that 5,000 years ago people were eating grasshoppers that had been washed up from the lake and dried in the sun. Experiments conducted to estimate the reliance on this kind of food source demonstrated that one person can collect 200 pounds of insects per hour. Cooked, medium-fat beef yields 1,240 calories per pound, while wheat flour yields 1,590 calories per pound. Astoundingly, insects come in at 1,365 calories per pound amounting to an average of 273,000 calories per hour of effort. A return of this kind is greater than the per

Table 1. Nutritional value of insects compared to typical foods

Type	Energy (Kcal)	Protein (g)	Vit A (ug/100g)	Vit B2	Vit C	Iron	Calcium	Phosphorus	Magnesium
Termite	613	20.4	2.89	1.98	3.41	27	21	136	0.15
Caterpillar	370	23	3.12	1.25	2.22	2.01	8.57	100.5	1.56
Grasshopper	611	26.8	6.82	0.07	8.64	1.96	42.16	131.2	8.21
Hamburger lean	140	25.9	0	0	0.11	3.9	14	233	0
Roast Chicken (3.5 oz)	239	27.3	47/161*	8.5	0.06	19	15	210	0.02
Baked sole	202	30	0	2.5	2	1.4	23	344	0

Source: Compiled from Banjjo et al. 2006; DeFoliart 1992; Pennington and Douglass 2005.

Note: All vitamins and minerals except vitamin A are measured in mg/100g.

Note: * RE/IU (Retinal Equivalents/International Units).

hour rate from seeds, which contributes about 300 to 1,000 calories per hour. Large game animals, like antelope or gazelle, provide only 25,000 calories per hour; hence, insects are certainly the most effective source of caloric intake; a return totally comparable to what a hunter might get as return on his game animals (McGrew 1994; McGrew et al. 2005). Some aspects of the nutritional value of insects compared to typical food animals based on a 100-gram serving are represented in table 1. This table indicates that, nutritionally speaking, insects make a good meal: they are higher in energy, equivalent in protein, and higher in most vitamins than typical foods. For the early hominin, then, meat—especially in the form of insects—became an increasingly important dietary adjunct, providing high-quality nutrients all in one package.

Although food may have been the primary reason for the Ancestor to approach fire, there are other attributes of fire to consider as well. Heat and light are the two most obvious features. Six million years ago, Africa was cooling, which is why the grasslands were expanding. The bipedal hominin *Orrorin tugenensis*, a candidate for the Ancestor, lived in the Tugen Hills of Kenya. Temperatures in the highlands are quite cool at night and used to average about 18°C (Alsop 2007) and, as discussed in chapter 4, fire would have provided comfort, as would light. Savanna chimpanzees in Senegal prefer to feed in the gallery forests that cut through their home range. It is cooler and there is more food. But they must cross open land to get there. Our Ancestor would have been faced with a similar problem. A small-bodied person—and the Ancestor was about the size of a female chimpanzee (Galik et al. 2004), that is, about 1.12 m (4 ft) tall—walking on two legs is vulnerable. There are felines that can outrun a biped; there are even eagles that can (and did!) pluck children and carry them off. Nighttime without trees to nest in would represent a particularly defenseless moment, but to cross grassland to get to thicker woodland might require camping out. Fire with its heat and light endowed the Ancestor with power and the ability to ward off danger.

All that was required of the Ancestor intellectually was to be able to throw things in attack or in defense—chimps and monkeys do—and to recognize fear in another creature. The latter would require a theory of mind: *I know that you know that I know. . . .* Monkeys and chimps both share this ability with humans, though to a lesser extent (Heyes 1998; Tomasello et al. 2003). Having at least the capabilities of contemporary monkeys and apes, the Ancestor would have had sufficient ability to know that being near a fire meant that other animals would not approach. What a boon a fire would be at that point. Gathered around a burning stump, mosquitoes are kept at bay, cats do not approach, nor do other denizens of the night. The chill is off the air. A moment of repose is experienced.

These three behaviors, then—remembering, eating charcoal, and insect-eating—give background for the Ancestor to seek out insects, to be familiar with burnt-over areas, and to remember what happens after a fire has passed. And the benefits of fire—heat, light, and protection from other animals that fear it—add immeasurably to its attraction. But if fires are also terrifying, how did nonhuman primates and the Ancestor overcome this fear?

The Roles of Social Context and Genetics in
Learning to Approach Fire

It is likely that the Ancestor shared with contemporary nonhuman primates the *circumstantial* fear of fire—an exquisitely primed system that is resilient and sensitive to information acquired from the cumulative learning of a group. Indeed, in the normal development of contemporary human children, fears develop in a particular sequence. In infancy, the basic fears are from environmental stimuli, like loud noises or loss of support. A little later, in the second six months, the baby begins to fear strangers, heights, strange objects, and separation. There is considerable cognitive maturation implied in this, as the child must be able to discriminate known from unknown. Fear of darkness and of animals (snakes, crawly things, etc.) begins to appear just around the preschool years (Gullone 2000). A child is a culture-bound animal from very early on; its language patterns are set by the singsong quality of its mother's baby talk. By the time the child is a few months old and has playmates, it is learning their values, fears, and desires; hence, what fears are "innate"—that is, hardwired in an organism (if any)—are difficult to assess. So how did cognitive development, social context, and genetic predisposition all come to have a role to play in mitigating the Ancestor's fear of fire?

Let us start with neurological processes. Memory formation is a dynamic process in which remodeling or reorganization joins new information to preexisting memories (Lee et al. 2008). A recent study describes how memories can change, even memories of fear. The process is biochemical and occurs on the far side of the juncture between neurons. Under experimental conditions with rodents, memory was disrupted by "degrading" proteins. New or altered memory was updated in a similar manner (Lee et al. 2008). In this way new information can "rebuild" memory, and what was fearful may lose its sting.

But the response to a primary stimulus is also influenced by the context in which it is received. Group-living animals and animals with caregivers have their responses profoundly influenced by the adults around them. Among human groups, the process by which the impulse to run from fire is dissociated from the image of flame and smoke depends on the reaction of whomever is holding the child at the time. If "gran" unintentionally tightens her hold on the baby when she passes the fire-place, or when the fire engine races by, alarms wailing, gran's reaction is felt and absorbed by the child. Adaptive behavior takes into consideration the fact that neurochemistry and genes do not work in a vacuum.

Our primate relatives are dependent on the information they get from living in a group—tried-and-true patterns that are passed down from one generation to another. This is true even of seemingly unlearned fears, such as the response to snakes or other poisonous denizens of the earth and trees. In the 1970s, the reactions of baboons to scorpions were tested. Those baboons raised within the group in their natural habitat would grab a scorpion, carefully pull out its stinger, and eat it with impunity. Those caged since birth or early youth, with little experience among other members of their group in their natural habitat, showed the same fear as usual: shunning it, vocalizing, and moving quickly away from it. The importance of certain animals in the environs of monkeys is demonstrated, for example, by the fact that vervets in the savannas of Kenya have distinct sounds for different kinds of danger: eagles and other predators in the sky, big cats on land, and even snakes in trees and bushes (Cheney and Seyfarth 1990). Do such vocalizations represent cultural adaptations or innate ones? The evidence suggests something more like culture. *Seeing* fire should link to the action of *fleeing*; this is called "dual coding" (Sadoski et al. 1991). However, since the response to a primary stimulus is influenced by the context in which it is received, a predisposition to fear—that is, a biochemical hook-up—is there, but what gets registered as fearful is apparently acquired via the social-cognitive processes of association and imitation. Interestingly enough, these same processes play a role in how fear gets extinguished (Phelps et al. 2004).

The process of *extinction* plays an important role in mitigating primate response to fire. Extinction in this context means that the perception of the original fear cue has been altered (Bouton 2002). Recently, it has been found that the anatomical mechanism in the brain by which this extinction process occurs is conservative; that is, it has remained similar over a wide range of animals throughout evolution (Phelps et al. 2004), and so it must have been similar in the Ancestor. Experimentation in the domestication of foxes demonstrated that fear and aggression toward humans was reduced due to a lessening of the activity of the pituitary-adrenal axis.

The actions of several hormones, in particular serotonin, were changed in these foxes, thereby altering the regulation of emotional-defensive responses, especially toward humans. This experiment is a "model" of hormonal action in reducing fear (Hare and Tomasello 2005). Note too that learning itself creates new junctures between nerve cells in the brain. The synapses, or neuronal connections, are not only inherited but, more importantly and significantly to this book, they are produced by experience (see Gilbert 2000, chapter 22).

In understanding how fear of fire became circumstantial it is useful to consider the relationship between image and emotion. Some fires are terrifying, such as those that are moving at tremendous speed and growing; others do not inspire as much fear, such as those that are limited to a lightning-struck tree, are seen at a distance, or where knowledge of the aftereffect contradicts the initial reaction. Here, the issue becomes the relationship between the *image* of something and the *emotion* generated by that image. In their discussion of the origin of ideas, Greenspan and Shanker (2004) take the relationship between image and emotion a step further. The image of something produces an emotion. If the emotion is reassociated with something else, something nonthreatening or even comforting, the *meaning* associated with the image changes. Greenspan and Shanker suggest that when an image is released from the usual and instantaneous reaction and instead is endowed with a different emotion, it takes on this new sense. This process is at the base of the development of ideas.

The relevance here is that the reaction of fear is attenuated, dampened, and replaced with calm comfort. As a result, a change in the biochemical substances that modulate emotion will also take place. Opioids—endorphins resembling opium—are involved in emotion, memory, learning, pain, and perception itself. Primates naturally produce opioids as part of their systems; these work together with prodynorphin, a protein building block for those opioid endorphins (Rockman et al. 2005). Prodynorphin is completely different in human and nonhuman primates. Whereas humans have at least five versions of prodynorphin, nonhuman primates have but one. Rockman and his colleagues suggest that this is a clear example of natural selection operating after the divergence that split us from the apes.

What Contemporary Observations of
Wild and Captive Chimps Can Teach Us

Is there anything to be learned from both wild and captive chimps about the Ancestor's potential willingness to approach fire? The story of Gracie from chapter 1, the cigarette-smoking chimp living at Holloman Air Force Base, instructs us that modern, nonhuman primates can learn rather astonishing and even quite sophisticated ways of engaging with fire. In addition, an enrichment experiment with captive chimps, conducted by Arnold Chamove in Gabon at the Medical Research Center of Franceville, allows us to consider further clues as to the capacity of the Ancestor to associate with fire. There, chimpanzees were housed in two enclosures, in the larger of which Chamove made a small fire of hardwood using sticks, which, for safety's sake, were no longer than 6 centimeters (Chamove 1996). The experiment lasted 4 days, and I cite here the full description of Day 1:

> [T]here was great excitement upon release, and the chimps circled the fire, keeping about four meters distance from the burning pile. After only about a minute, the dominant male ran past, hitting the pile of burning wood with his hand and scattering the burning sticks. A few of the animals then approached to less than a meter distance, inspecting the sticks. After about two minutes, one chimp picked up a stick, sniffed it, and dropped it. He then picked up a piece of burned charcoal that had broken off and ate it. Three animals then hit at the burning sticks. After three minutes, there were still many pant-hoots and much excitement, and dominance displays were shown. At four minutes, a female used a cold stick to poke at and break up the fire. At five minutes, a female chimpanzee raked together smoldering embers, another hit at a smoking stick and a male had an erection as he hit at a cold stick with his hand. The rest carried away sticks and chewed them, leaving nothing at the original fire location.

Over the next 3 days, the excitement gradually eased until, at the end of the experiment, the chimps virtually ignored the fire. However, on

Day 2, after 7 minutes of exposure, the dominant male removed a stick by means of its unburned end and then placed it back into the burning pile. On the fourth and final day, chimps actually retrieved and ate burned sugar cane and banana, although they did not put anything into the fire (Chamove 1996).

Jane Goodall is quoted as noting that chimps do not seem to be afraid of fire (Clark and Harris 1985). Bill McGrew, an observer of chimpanzees for decades, quotes a colleague who saw chimps in Senegal managing campfires "for cooking and warmth," as well as extracting seeds and nuts from bush fires (McGrew 1989). There is a dearth of other published reports on ape behavior toward fire, although there are anecdotal reports of orangutans "playing" with fire in Kalimantan and gorillas filmed approaching fire with sticks. The documented findings are as incredible as they are suggestive. Here are animals, qualitatively different from other mammals, able not only to interact with fire, but to utilize it. These apes managed fire, put wood on it, and ate foods that had fallen (or been put) into the fire. Scholars have noted that ape behavior in captivity is often more sophisticated and more suggestive of human behavior than is their behavior in the wild. Certainly the use of their innate intelligence shifts from assessing the myriad of stimuli around and about them to exploring those near at hand in more detail. That apes in the wild—that is, beyond captivity and beyond human example—do not do what captive or rehabilitated ones do is important, but does not change the fact that they are *capable* of doing such things, given a particular context.

The simplest way that new behaviors become part of a group's repertoire, whether we are speaking of human or nonhuman primates, is through aging. Juvenile innovations become included by dint of the innovator's repeated use of the pattern and continued use as the juvenile becomes older. Youngsters younger than the innovator replicate the pattern because the older member is a model. By the time the juvenile is an adult member of the group, and assuming a position of sufficient prominence within the group, younger members will imitate the pattern because "that's the way it's always been done" (Burton and Bick 1972). Hence, picking up burning sticks might quite readily have become part of what the Ancestor's group did, granting this Ancestor only ape intelligence and a little more, especially given the significant genetic differences between humans and apes. The larger issue of culture and its relationship to the likely path of our Ancestors' permanent acquisition of fire technology is detailed in chapter 5.

What, then, is the importance of these modern observations of non-human primate behavior to speculation about the Ancestor's earliest approach to fire? Perhaps these data show us the possibility that somewhere in the millions of years of anthropoid evolution prior to the divergence with apes the initial step in accommodating to fire had already been taken. As observed earlier with baboons, the flight response to the fire stimulus was weakened, and the inclination to perceive flames, smoke, and heat as objects of danger was attenuated by utility or adaptation. Overriding the fear response was curiosity and, as suggested by the chimp experiment, excitement—excitement very much like the behavior described by Jane Goodall in the "rain dance," where chimpanzees respond to elevated excitement by picking things up and throwing them, vocalizing, crashing about, and piloerection (hair standing up). This is not a fear response, but may be closely allied in the autonomic nervous system. Between excitement and play behavior is the experience of being near fire without negative results—feeling the heat and seeing the light without terror of burning. This information is also stored. Perhaps the group is attracted to a fire because they know from past experience that insects are abundant in the aftermath of a fire. Arriving at the fire site, there is general excitement; the adult males whoop and grab sticks, the females with infants retreat to a safe distance, while the juveniles start crashing about, flinging branches, grabbing lit sticks from the fire and flinging them, too.

What the Ancestor Was Learning about Fire

Learning from observation or experience without control over the thing being observed or experienced is not science, but it may be considered a precursor to it. These observations and experiences are quite powerful and were undoubtedly a source of knowledge for our Ancestors as much as for our contemporary primate cousins. Since the ability to predict fundamental environmental phenomena is necessary in order to live in a particular environment, it is my belief that, for the Ancestor, proximity to fire through association with cooling embers was at the beginning of the technological revolution in which fire was tamed. We know that monkeys and apes adore termites and other insects and that fire-loving pyrophilous insects approach fire-burned areas. I speculate that, like birds of prey that fly *toward* a fire front looking for animals running before it, the Ancestor, a small creature by all accounts, would have approached areas where fire was burning to await the arrival of dinner. Those chimps that do gather insects can be extremely creative—the tools that some groups of chimps have invented to procure termites are sophisticated and complex. In the dry habitats of Senegal, for example, a variety of woody or viny-supple vegetation is used (McGrew et al. 2005), but in Congo a more sophisticated set of implements specific to the task have recently been found. There is a puncturing stick used to excavate a hole in the termite mound. Then a twig, carefully and skillfully shredded at its tip to make a brushlike object, is inserted into the mound, its complicated surface allowing more termites to cling to it and be extracted (Sanz et al. 2004). The kit is specific to location and task, and, what is more, the sources of the tools are chosen for their attributes such as strength, or flexibility: the puncturing stick comes from a tree (*Thomandersia hensii*), but the brush-tipped probing tool comes from a species of herb belonging to the genus *Sarcophyrnium*. The fact that the chimps seek materials from particular sources and then fashion each tool differently and according to purpose speaks volumes (Sanz et al. 2004).

With very basic capabilities, the Ancestor would have quickly learned that the fuel used to feed the fire had to be dry. The Ancestor would have seen, but perhaps not considered, the water vapor that escapes in the form of steam as wood dries and flames. Despite a lack of formal knowledge about fire, the Ancestor would have related to fire's basic properties as we do. Hence, fire would have been appreciated for its ability to provide heat and light. That the heat and light increased as the fire grew bigger would have been recognized. The Ancestor would have realized that a suffocated fire will no longer burn—and this in the absence of any knowledge of oxygen.

Even before formal science, humans understood the relationship between the generative light of the sun and the symmetry with burning: the heat, the light, the colors. While the Ancestor may have thought these to be the same, we see a difference. We know that the building process we call photosynthesis occurs when plants take light and use it to convert carbon dioxide and water to be stored as sugar, and we are familiar with the breakdown process of burning whereby heat and light are once again released. We now readily understand energy exchange and sunlight as photons—that the process of manufacturing sugars in plants is a transformation of light itself by use of that energy.

The Ancestor could not help but notice the comforting smell that is characteristic of burning wood, but would not have known that it came from lignin, the major constituent in the cell walls of wood. That fire produces smoke would also have been noticed; however, there would have been no knowledge of the fact that smoke is composed of small particles suspended in air, the result of incomplete combustion of the wood fuel (Harris 1980). And most certainly, the Ancestor would have noticed that insects avoid smoke. Hence, to the properties of fire that the Ancestor recognized—that is, heat and light—must be added smoke. Smoke in the distance heralded a fire, signifying either the presence of others nearby or a natural burn, important if the source of the Ancestor's fire had been extinguished.

Learning to Predict the Location of Fire

The Ancestor could find and predict the *location* of types of fire that could be of use, and they would have noticed the *causes* of fire—rocks colliding during landslides, producing sparks sufficient to ignite a grass fire, volcanoes spewing hot embers, and above all, lightning striking dry vegetation, were all dramatic sources. There were also events less dramatic than these that provided fire. The Ancestor would have been familiar with sources of natural fires, including, at least in parts of South Africa, coal seams and peat fires. Spontaneous ignition of coal continues in this area, the fifth largest producer of coal in the world. The toxicity of the area where these organic sources are burning indicates an inhospitable zone for any organism, but may have presented known locations for sources of fire. Peat bogs, just recently studied in Mali, ignite spontaneously, especially under conditions of drought, and, when conditions are dry, that combustion is as likely as not to be the result of bacterial activity. Self-ignition occurs as organic-rich layers of material degrade biologically, similar to the process of decomposition and heat production in a compost pile. Surface temperatures can reach 765°C (Svensen and Dysthe 2003).

The Ancestor would have known the *conditions* under which the probability of a fire igniting would have increased: dry material, dry air, and hot ambient temperatures. The consequences of a wildfire would have been remembered, including the fact that certain plants only grow after the heat and smoke of the fire have opened their seeds (Read and Bellairs 2003) or the fact that a shift occurs in certain animal populations—some increase in density while others disappear. The Ancestor would have been able to predict the direction of fire to some extent. Wind factors and topography play a complex role in how a fire progresses, so predictions would have been limited but nonetheless helpful in guessing what tomorrow might bring. This scenario is not far-fetched. Snakes and other burrow dwellers can predict earthquakes, probably from the vibrations in the ground, and monkeys in Gibraltar can predict the coming of the Levant—that cold, dank weather condition much dreaded for its negative effect on health—and will stay for days where they find shelter, out of the way of the inclement weather. It is not too fanciful, therefore, to picture the Ancestor looking up at the clouds, sniffing the air, and feeling the change in pressure that precedes a storm. The storm would surely have caused lightning, the lightning would surely have struck, the strike would have caused a burn, and the burn would have been a source of fire.

A recent study estimates that there are eight million lightning strikes every day, resulting in eight billion tons of burnt vegetation each year (Scott 2000a, 2000b). Fires started by lightning have a tremendous impact, are truly awe-inspiring, and must have been regarded by the Ancestor as a serious threat to the surroundings, including to other animals known to be living in the same environs. Thunderstorms accompany some, but not all, lightning strikes, and not all lightning strikes cause fire (Whelan 1995). If ambient temperature is low and leaf litter is wet, lightning has no ability to ignite vegetation. Unstable summer conditions, when air masses heat up over hot, dry land, are most likely to see lightning cause fire. Volcanic activity, within the lifetime of the Ancestor, also played an important role in igniting fires. Just beyond our time frame, and nearly 4 mya, comes evidence that bipedal ancestors ventured within some

20 km (12 mi) of a volcano, close enough to fossilize evidence of their promenade (Palmer et al. 2005). Imagine, if you will, a time shortly after a minor eruption, the ash cooling and accumulating, the burning trees providing a source of fire. Some chimpanzee-sized people walk along— perhaps they are looking for a branch they can obtain, or perhaps they are just wandering. There are footprints that date from 3.7 mya from Laetoli, Tanzania, that touchingly show what might have been a man and woman, and perhaps a third person, walking together through grassland savanna and leaving imprints filled in with volcanic ash.

Learning about the Properties of Fuel

I have not decided whether the Ancestor, like contemporary humans, was opportunistic in picking up twigs, sticks, and branches to nurture controlled fires, or whether wood was specifically chosen based on certain inherent burn properties (Asouti and Austin 2004). Both choice and opportunism existed in the ethnographic present (that is, until 1965), and even chimpanzees choose certain woods over others for their termite hunting (Sanz et al. 2004). The chimpanzees' tool for puncturing a termite nest, as previously mentioned, comes from one specific tree, and their probes invariably from a specific herb. It seems that the chimps are aware of the properties of each kind of plant and which plant is particularly well suited for which activity. What is more, these particular raw materials are sufficiently important that chimps have been seen traveling a distance to obtain them (Sanz et al. 2004), implying not only memory, but memory of location with specific reference to a particular object. I grant the Ancestor at least this capacity.

Although all wood is made of carbon, hydrogen, oxygen, and nitrogen, as well as water, gases, oils, minerals, and other compounds, the proportions differ. As a result, the heat available differs between softwoods and hardwoods, which in turn means that some woods have higher kindling temperatures than others. Of the basic elements comprising wood, lignin, the structural protein of plants, is the most abundant. Furthermore, since there is more lignin in softwood, there is also more carbon and resin than in hardwood, which is why softwood produces more energy than hardwood (Cheremisinoff 1980). This seems counterintuitive: everyone ever making a campfire prefers hardwood for the "better" burn, but while hardwood will give approximately 20,000 kJ per kilo, softwood produces almost 2,000 kJ more. Yet a hardwood log (oak) with the same dimensions as a softwood log (pine) will weigh more because it is denser. So per *cord* the hardwood will give a better yield (Helfferich 1996). Bark, especially bark from softwoods (Tillman 1978), is easy to gather from dead wood and fallen trees and has a high heating value, as well. The average energy output is about 8,280 Btu/lb (4,603 kcal/kg).

Clearly, with a simpler cultural base, the Ancestor would not have needed wood for the same variety of purposes for which modern peoples use it and for which specific woods are chosen. Different functions require different qualities to be present in the wood: a bow needs more spring, a canoe less porosity. Historically, special woods have been chosen by various peoples for such things as dugout canoes, masks, weapons, medicines, or certain kinds of construction. The Ashanti (central Ghana), in their rituals supporting the choice of chief, utilized a tree called "kill the python," of which a certain number of cuttings were to be put in specific places in a particular pattern. In a spirit-possession ceremony, the head of a Bunyoro (Cameroon) household had to collect the wood from an engando tree because the light it gives off is exceptionally bright. In the mid-nineteenth century, Sir Richard Burton wrote extensively and interestingly of his travels in Africa. He described the medicinal use of wood in Somalia in which two pieces, one shorter than the other, were selected to be used to heal him based on their dryness and medicinal power. The longer piece was placed into a scooped-out hole in the shorter and rubbed until it smoked and charred, at which point Burton's stomach was cauterized. Clearly, he suffered no ill effects from the treatment as he lived to tell about it. Needless to say, wood is also specifically chosen for more prosaic activities. Some Ibo (Nigeria) prefer to use coconut and a kind of palm as house posts, while bamboo is employed specifically for rafters. The Mbuti Pygmies of the Ituri Forest (the Democratic Republic of Congo) carefully select wood and leaves in order to maximize the amount of smoke they will produce when they go to gather honey. In Africa and India, species of acacia trees are currently favored as firewood. Some species, like *A. tortilis*, are preferred sources of firewood for making quality charcoal in Tanzania and the Sahel because they produce prodigious amounts of heat, reckoned as 4,400 kcal/kg (FAO 1995).

These samples of a precise fit between a given type of wood and the use to which it is put indicate the depth of knowledge people can attain when they come to depend on a given material. I doubt, however, that the Ancestor, when initiating this relationship with fire, gave much thought to such details, although we must assume a degree of forethought at least equal to chimpanzees working their termite mounds. More likely, as much an opportunist as an empiricist, the Ancestor took whatever appeared dry and used it to keep a fire found burning on a stump, or husbanded from some days ago, from going out.

Footage produced by the Australian government of the Aranda from before traditional ways were supplanted shows just that. One of these films depicts a family moving from one temporary site to another. The father leaves first, carrying his firestick. He keeps this ignited by use of burning grass as the family moves through desert, the mother, or mothers and younger children together, older children on their own. After finding a spot to spend the night, preferably not too far from water and well away from prowling animals, the family congregates, and the firestick is used to ignite the dry twigs, bark, and small branches gathered by the family as it moved along. No structures are built at this time; the family sleeps on the ground with nothing but leaves as bedding. It is cold in the desert at night, so the family places itself around the fire in order that at least one side is warmed as they sleep.

The anthropologist Elizabeth Marshall Thomas (1959, 210) described a night with the San of the Kalahari in these terms:

> That night it was so cold that no one wanted to leave the fire
> and when finally everyone was at my camp, settled close to
> each other, side by side, with their bellies warmed by the heat
> and their shoulders covered by their capes, their backs turned
> against the enormous veldt where the cold wind blew and
> where the dark was coming, they began to entertain themselves
> with conversation, their favorite pastime. The adults held their
> elbows on their raised knees and cupped their hands under
> their chins to cast little shadows into their eyes, for in this way

they could see clearly over the bright fire. Close to the fire even the children were not cold. The older children sat up like the adults, although they were naked and the adults had capes, and the younger children slept in their parents' arms, warm between the fire and their parents' skin. In the morning, in . . . the first light, we woke up . . . [to] a fresh, warm wind blowing. It was easy to sleep now, and the people slept for hours, rolling away from their piles of ashes as the day warmed.

For the Ancestor, dependence on wildfires would have grown as this new instrument proved increasingly valuable. The Ancestor would have noticed how wildfires preheat vegetation as they move forward. A sort of chain reaction begins as the plant material at the van of the fire warms and dries to the point of combustion (Vogel 2003). The chain ends with the remaining charcoal smoldering and burning, leaving only ash.

Categorizing and Nurturing Fire

Like contemporary foresters, the Ancestor probably had criteria with which to categorize fires. After all, as discovered by Ernst Mayr, the noted evolutionary biologist, peoples in New Guinea used a system of classification for their plants that uncannily reproduced the scientific, Linnaean one. Nonhuman primates categorize phenomena in their world: plants, birds, other individuals, and more. They know dangerous from innocuous, edible from inedible, and friend from foe (Burton 1984). A functional evaluation based on observation was important to someone living in the savanna, much as the habitat of the Innuit motivated them to recognize different qualities of snow. The most obvious kind of fire was grass fire, moving quickly along the surface of the earth, consuming shrubs, grass, and the lower branches of trees. Not as obvious, but perhaps more significant, would have been ground fires, burning close to the ground or even within it, running along roots or even in organic soils. Their importance, if it was recognized, was in being the source of new fires when the conditions were right. Not as terrifying as a fast-moving wall of flame, this kind of burn could have been anticipated by the Ancestor, making it accessible. Crown fires would have been more spectacular, as they jumped from tree to tree, dislodging large embers and reaching incredible intensities. Fire would have become a treasured commodity, something that became indispensable in the way modern technology is to us, in an incredibly short period of time.

Long before being able to make fire, the Ancestor would have nurtured it. A nearly contemporary model suggests what *might* have been. North American pioneer women left the natal home with a sourdough starter, so that she might bake bread as her mother had done, and embers from the family hearth to comfort and warm her new home. Clark and Harris (1985) report a similar practice in India, where tribal peoples carry charcoal embers from one home to another and feed them until they reignite at the new hearth. This transfer of sacred fire from one home to another has been found in a multitude of cultures around the world and over time. The Nuer (Sudan) have stories in which the plot centers around a woman going to the home of another to fetch fire (Hutchinson 2003). The Masai of East Africa make fire by having one person hold the fireboard and another manipulate the drill stick. Once the chips of wood have caught and fire is established in that hut, neighbors come from other huts holding sticks of firewood in their hands to fetch fire back to their own homes (Talbot 2003).

At first, then, the Ancestor probably only sought fire out and remained with it until it self-extinguished. Perhaps, as archaeologists have suggested, the Ancestor approached tree stumps that were smoldering and stayed by them. Clark and Harris (1985) reflect on a number of peoples in India, as well as Africa, who do just this. Indeed, old, dead, or lightning-struck trees would have provided a virtual torch in the ground.

Intensifying the Association:
The Beginning of Control over Nature

Furthering an association with fire—and choosing to do so—is a cognitive process that differentiates human from ape. The distinction intensifies with the next cognitive jump: feeding the fire. This action marks a shift from passive recipient to active agent, from beneficiary to manager. It is the beginning of humanity's control over nature, and it probably went unnoticed by the Ancestor. At the same time comes the extension of known chimpanzee behavior—touching, holding, or even flinging a stick on a fire does not seem to be beyond the Ancestor's capabilities. And yet, the act of picking up a lighted bough distinguishes the Ancestor at the time of the divergence from all other animals. It is the material of legend and religion; it is the moment of radical change. Being willing to hold a firebrand, nurturing the campfire, carrying glowing embers—these behaviors describe the set of acts that fixed humanity's dependence on controlling energy. We might place them within a sequential model as follows:

1. Approach and associate
   a. food
   b. protection (warmth, light)

2. Nurture
   a. dried grass
   b. twigs

3. Manufacture—the invention of *Homo sapiens* (sometime after 500 kya)

The argument of this chapter has been that, given capabilities common to Old World monkeys and apes, the Ancestor, with a brain no larger than a contemporary ape, but who was a committed biped, began a relationship with fire hitherto unknown in the animal world. The motivation to do so may have originally been an easy and ready source of food, but soon grew into dependence upon the light, heat, and protection provided by fire. Bringing fuel to feed the fire would have been a natural outgrowth of observation spurred on by this dependence. There is a progressive growth in the control of fire, and there is a metaphorical importance inherent in it. What, after all, could be more fearsome than a two-legged beast brandishing a stick hissing with fire and blinding with flame? What stronger act of defiance could there have been? What more indelible memory of humanity's place in nature could our Ancestors bring down the ages than of themselves overpowering those that would have made them dinner?

Let There Be Light

Altering Hominin Physiology by Extending the Day

The more frequent and regular contact with fire would surely have had physiological consequences for the Ancestor. In this chapter I explore whether the light of the campfire would have had the effect of sufficiently "extending the day" for this Ancestor, with the implication that a biological phase shift would have occurred, increasing the body's "daytime" by decreasing the hormone melatonin. Because melatonin in turn affects many other physiological systems, in particular patterns of brain activity and reproduction, the acquisition of fire and its broad influences on the "biological clocks" of the earliest Ancestor could in fact have been the most critical factor in explaining the divergence of our species from kindred apes. What is the scientific underpinning of this line of conjecture, and what are its implications for human evolution?

Life on this planet depends on sunlight, and organisms have evolved to be able to use this light as the means to coordinate internal rhythms with the environment. All forms of life, from the tiniest single-celled creature to humans, react to light-dark sequences and experience cycles in the fluctuating amounts of molecules—photosensitive proteins—in their cells. There would be absolute chaos in an organism due to the complexity of regulating the increases and decreases of molecular switches except for the pacemakers that regulate networks of physiological rhythms. This is accomplished by means of external, environmental cues, particularly the daily and yearly photoperiod.

Photoperiodicity—the response of an organism to cycles of dark and light—is a function of the rotation of the earth around the sun, which dictates the amount of light falling on a given area at a given time. The variations in light—the greatest amount or least amount of light—reach their peak at the equinoxes due to the earth's position relative to the sun as it travels in a figure-eight-shaped orbit. The earth is tilted on its own axis with reference to this ellipsis at 23.5°.

Life forms have utilized this predictability to create the intricate patterns of activity, growth, and reproduction that we observe (Kuller 2002). Photosensitive protein molecules react to environmental cues to set these networks of physiological rhythms. Life-history cycles, like sleep-wake, the rise and fall of body temperature, or reproductive states, oscillate and are commonly referred to as "clocks." In all organisms these clocks control all the basic functions: biochemical, physiological, and behavioral (Sancar 2000). Such circadian clocks ("circadian" referring to a 24-hour interval, or "about a day") occur throughout the body, but require daily "enforcement" (Buijs et al. 2003) from the one master clock—the suprachiasmatic nucleus of the hypothalamus, which both generates and synchronizes the oscillations (Stehle et al. 2003). For example, blood cells are "refreshed" from within marrow by stem cells that respond to circadian rhythms. In nocturnal mice, they peak 5 hours after light and reach the lowest point 5 hours after darkness (Méndez-Ferrer et al. 2008). The suprachiasmatic nucleus, sitting just above the optic nerve in the hypothalamus, has its own independent "pacemaker" clock cells that are under genetic influence (Stehle et al. 2003). The role of the suprachiasmatic nucleus is implemented not only by hormones, but also by direct neuronal action on organs of the body. It therefore regulates daily well-being, preparing the body for the anticipated period of activity by raising heart rate, sugar, and stress hormone (cortisol).

While experimentation shows that circadian rhythms are internal (endogenous) in an organism—in effect, hardwired—it is sunrise and sunset that are the triggers synchronizing personal periods with the actual day. How exactly do these triggers work? We know that cells of the retinal layers of the eye convert light into an image so that we "see." Less well known is that in addition to visual cells, the retina contains special nonvisual photoreceptors that convert light stimuli to chemical substances, which themselves affect a variety of target organs. It is this system that regulates our circadian rhythms.

The technical term for the agent that entrains such cycles is "time giver," which sounds even more picturesque in its original German—*Zeitgeber*. The light-dark cycle is the major *Zeitgeber*. A 24-hour oscillation has *phases* within it. Dawn and dusk, for example, act as reference points defining a 24-hour phase. When an organism is kept in darkness for a period of time, its endogenous clock will "free run" on an almost 24-hour cycle. Mammals for example, have a personal period of around 23.95 to 24.10 hours (Gorman and Lee 2002). Light will shift the period by an increment each day until the phase approximates 24 hours. Phase shifting is based on the genetics of an organism and its response to environmental cues (Aton et al. 2004; Green and Besharse 2004). *Reset* is the term used to describe such phase shifting because the clock has a periodicity that is being adapted to a particular schedule. An internal reference would be something like a rhythmic secretion of a protein or the beginning of some activity.

Until the mid-1990s, it was thought that the intensity of light on the visual receptors was the source of circadian rhythms. Now the receptors, pathways, and target organs are identified, and the interpreters are known to be the pigment melanopsin—which leaves specific cells in the retina to travel into the brain, where, through a circuitous path, it finally ends up in the pineal gland—and *cryptochrome*, which appears to be an essential photoreceptor in humans as well as other mammals, working in concert with melanopsin (Sancar 2000; Cashmore 2003; Sancar 2004). Cryptochrome is a key clock protein, mediating circadian rhythms. It is regulated by two genes named *CRY1* and *CRY2* (Sancar 2004) and—at least in rats—is directly under the control of melatonin in one part (the *pars tuberalis*) of the pituitary (Dardente et al. 2003). In addition, three other genes—*DEC1*, *PER1*, and PER2—each activate at a different part of the light cycle. It is a marvel of science that the system has been decoded.

What makes it even more complex is the discovery of orexin (hypocretin), a neuropeptide found in retinal cells whose function is hunger and alertness (Bazar et al. 2004). The evolutionary significance of this is clear: there is at least one redundant system that ensures that the animal will be alert and able to find its food.

The Role of Melatonin in Biological Rhythms

Melatonin, the hormone that plays the major role in the transduction of information about the environment into biological rhythms, is synthesized from circulating tryptophan, which is then converted to 5-hydroxytryptophan, then to serotonin, and on to n-acetylserotonin, and finally to melatonin (Borjigin et al. 1999). A demonstration in the early 1960s of the daily changes in serotonin concentration showcased an important finding leading to the understanding of the synthesis of melatonin. In recent decades serotonin has become well known as a medication to relieve various forms of depression. While levels of serotonin can be increased by such activities as ingesting carbohydrates or smoking cigarettes, *melatonin* is insulated from any influences. Just a few years ago, the current view was that this singularity had evolved to prevent any other system in the body from manufacturing melatonin (Ganguly et al. 2002). As research continues, however, melatonin has been found to be synthesized in skin cells (Fischer et al. 2006), hair follicles (Fischer et al. 2008), and lymphocytes (Carrillo-Vico et al. 2004). These secretions add to the amount of circulating melatonin. An experiment on the strength and effectiveness of melatonin showed that administration of a dose as low as 0.1 mg could make subjects fall asleep about 10 minutes faster than usual and increased the duration of sleep by about the same amount of time (Brainard et al. 2001).

Melatonin is an antioxidant, directly scavenging free radicals and considered by some to be more powerful than vitamin E or even blueberries or pomegranates! Since melatonin enhances the production of substances like interleukin-4 in bone marrow, it influences immune responses (Liebmann et al. 1997; Méndez-Ferrer et al. 2008) and has potential for helping to treat cancer, attenuate aging, and influence other conditions such as lowering "bad" cholesterol levels in human women (Tamura et al. 2008), in addition to its role in mediating environmental information through daily and yearly cycles (Irmak et al. 2005). Melatonin may also play an important role in the regulation of hair growth cycles (Kobayashi et al. 2005) and can suppress ultraviolet damage to skin and eyes (Iżykowska et al. 2008). This raises the interesting notion that the evolution of loss of pelage, and the evolution in diversity of skin color, can be associated with the actions of melatonin (Slominski et al. 2005). All our physiology, it would seem, is ultimately related to the path of the sun creating the daily variations in light.

In addition, melatonin is the only known hormone secreted by the pineal gland, or more precisely, secreted in the cells of the pineal gland. Melatonin can easily diffuse across the blood-brain barrier. This membrane protects the brain by selecting what can enter from the blood. A variety of hormones throughout the body are influenced by melatonin via the pituitary, the "master gland" of the body. It is a pea-sized organ, sitting at the base of the brain, and controls all the other endocrine glands. Consisting of three lobes, the pituitary secretes hormones vital to the functioning of growth and reproduction, pigmentation, and water balance. The pituitary is sensitive to the action of melatonin. For example, experimental studies have shown that if there is a suppression of melatonin there will be a simultaneous increase in the pituitary hormone prolactin due to suppression of the gene *Period 1* (Lincoln and Richardson 1998; von Gall et al. 2002).

Photometry: Measuring the Psychophysical Properties of the Visible Spectrum

Visible light is a tiny part of electromagnetic radiation, ranging on the spectrum from 400 nanometers to 700 nanometers (see figure 2). Light is measured in several different ways depending on perspective and function. In terms of our interest in evaluating the nature of campfire light and its potential impact on the hominin brain, it is photometry that measures the sensation and perception of light by living organisms. All light measurement in the photometric system is defined with respect to direction

Table 2. Units of measurement			
Unit	Name	Abbrev.	Measures
Luminous flux	Lumen from candela	Lm	Flow from source
Illuminance (area)	Lux lm/m²	Lx	Reaching a surface
Luminous intensity	Candela	Cd	Intensity, strength
Luminance (area)	Nit=cd/m²	Nit	Brightness

and surface. The light-energy source, as it radiates outward in space, is described as flux. Its strength or power is described as luminous intensity, while its brightness is luminance (see table 2). Measuring from the source of the light is one thing, but measuring at the surface it hits is quite another. The units of measurement define these variables in precise terms. Of the various measurements that are taken, the one that has traditionally been used with regard to the human eye is lux. Lux measures illuminance—which describes the intensity of light falling *onto* a surface. This note on measurement of the nature of campfire light and its potential impact on the hominin brain utilizes traditional photometric measurements, despite the fact that the circadian system operates with different receptors than does vision. There are two different sites in the brain where these two different functions (vision and circadian rhythms) are processed: the former in the cortex and the latter in the midbrain, at the suprachiasmatic nucleus (Sancar 2004). So although photometry is still used in studies of biological rhythms, it may actually be of limited value (Rea 2002).

The Eye's Mediation of the Visible Spectrum

The visible spectrum arrives at the human brain through the extraordinary structure that is the eye (see figure 3). At the back of the eye is the retina, a multilayered structure in which are embedded the photoreceptors, including the circadian photoreceptors, and their visual pigments. It is where light is converted—transduced—from photon to chemical signal

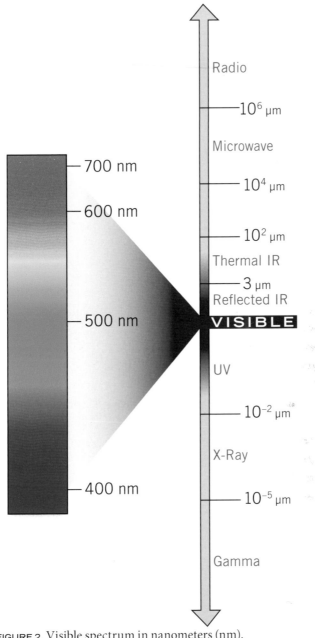

FIGURE 2. Visible spectrum in nanometers (nm).
The electromagnetic spectrum is in micrometers (μm).
Diagram by Kathleen Sparkes, after NASA diagram.

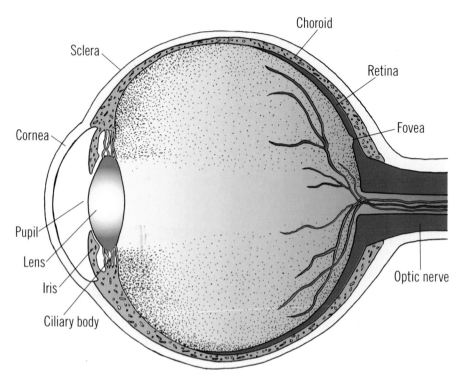

FIGURE 3. The human eye. Illustration by Mary Sundstrom.

to be sent out the optic nerve. Running out of the back of the eye to the brain are the branches, the axons, of the ganglionic cells of the retina, which form the optic nerve.

Within the retina are two critical regions where the layers are not present: the fovea and the optic disc. This complete lack of layers ensures that the photoreceptors receive light unimpeded by tissue. The fovea is at the center of the retina and defined by a line, the visual axis, drawn from the cornea to the retina. This is where vision is sharpest and where there are the most daylight photoreceptors—cones—numbering up to 7 million. At about 20 degrees from the fovea is the blind optic disc, where branches (axons) exiting from the retinal nerve cells join to form the optic nerve. The optic disc is blind because there are no visual photoreceptors in this region.

Visual photoreceptors are referred to as rods and cones. Their anatomy and function is complex, but suffice it to say that they each have an inner and outer segment, among other structures. It is the outer segment, near the cell nucleus, where the visual pigment is produced. Rods operate in night, or *scotopic*, vision and are sensitive below 400 nm. The human eye contains between 110 and 130 million rods. They are so sensitive that they are capable of picking up a single photon, which they transmit in black and white (Kalloniatis and Luu 2003). While both rods and cones are extremely sensitive, their sensitivity is due to different factors. Whereas rods contain more visual pigment than do cones, each cone is attached to its own nerve cell (multiple rods share one nerve cell). There is overlap in *mesopic* vision at twilight and dawn, when it is neither full day nor full night, and in moonlight. In this low-light condition, both types of visual receptors are operative. The mesopic period is of utmost interest to our story, as firelight produces about the same amount of light as twilight (10 lux).

The eye is rather like the light meter in a camera in the way that it senses lighting around the organism (Van Gelder 2003). The photoreceptors in the eye are made up of a kind of protein known as apoproteins and contain one or more photopigments, which are also proteins. Photopigments are molecules that absorb light near the ultraviolet-visible spectrum and are capable of converting light energy into chemical or neuronal information. The absorption of light initiates a chemical reaction whereby photons are transduced into nonvisual messages. This unique system relies on two structures: the suprachiasmatic nucleus and the pineal gland. The suprachiasmatic nucleus, with its tens of thousands of nerve cells, lies right above the brain stem, above the optic nerves. These nerve cells constantly monitor information that they then transmit to receptors in the brain and body. The information being monitored, in particular, is light. When light enters the eye, photoreceptors in the retina are triggered and the information is then conveyed to the brain. An image is reconstructed, first in the lateral geniculate nucleus (LGN) located in the thalamus of the forebrain. There is a "map" in the lateral geniculate nucleus that reflects each point on the retina. Signals from the retinal cells go to different layers in this structure depending on whether the eye through which the light entered is on the same side of the brain (ipsilateral) or the opposite side (contralateral). Information then travels from the LGN to the visual (striate) cortex of the cerebral hemisphere, such that right fibers go to the left of a particular fold in the cerebral hemisphere (calcarine sulcus), left fibers to the right of it, bottom fibers above it (occipital pole), and top fibers below it (CALnet 2004). This is how we register light as such.

The process whereby photons become chemical signals, also known as *transduction*, is how the energy information on one plane of existence is brought into or converted onto another plane. Photons—bits of light energy—are transduced in the visual photoreceptors by the visual pigment rhodopsin, itself composed of opsin and a molecule related to vitamin A (chromopore). Until recently it was assumed that chromopores were the molecules responsible for transduction of light in the nonvisual (hormonal) pathway as well as in the visual pathway. In the past few years, however, a body of research has emerged that has changed this view. First, another opsin, melanopsin, was identified as the key agent (Brainard et al. 2001; Berson 2003). More recently, cryptochromes have been found to work with melanopsin in the inner retina, where they receive the light signal and transmit it by means of specialized retinal ganglion cells to the master circadian clock, the suprachiasmatic nucleus (Sancar 2004). These "giant" ganglion cells are found uniquely in the *primate* retina—not in mammals generally (Dacey et al. 2005). The fact that it is found in primates suggests a departure a long time ago from the general mammalian pattern.

All of this information is new and makes history of yesterday's discoveries. That the visual pathway is different for circadian photoreception was not accepted until the late 1990s because it was so contrary to accepted scientific theory. As with so many discoveries in science, C. A. Czeisler, the scientist responsible for this work, could not get published because his studies did not conform to the consensual knowledge of the time. He and his colleagues studied blind people and found that their sleep-wake cycles were undisturbed, signifying that their circadian rhythms must be intact. This could only be true if vision and photoreception were separate functions (Czeisler et al. 1995; Lockley et al. 2003). The discovery meant that there had to be a unique photoreceptor that transduced photons into nonvisual messages. The importance of this finding cannot be underestimated—it suggests a unique system in the human eye that provides the brain with information regarding light but has a completely separate function from vision (Berson 2003).

Photoreception itself has a long history. Chemicals sensitive to light are known among some of the most ancient organisms on earth. The development of an organ with photoreceptors has taken hundreds of millennia and is found in anciently evolved creatures like octopuses and their relatives, as well as in mammals (Plachetzki et al. 2005). In 2006, James Bellingham and colleagues discovered that vertebrates have evolved *two* different genes that produce melanopsin. Nonmammalian vertebrates have both kinds of genes but mammals have lost one of these two types and function only with the other (Bellingham et al. 2006). The authors of this study suggest that this change may have accompanied a nocturnal episode in mammalian evolution. Certainly most of the contemporary types of the earliest primates (prosimians) are indeed nocturnal.

All primates can see color, but their ability to do so was a tradeoff, like the loss of the ability to synthesize vitamins B12 and C. Evolution is opportunistic—genes once functioning to develop one organ shift their work to develop another. Structures functioning in one way are therefore said to become appropriated for another job. The vomeronasal, also known as Jacobsen's organ, is a case in point. An ancillary organ of smell in mammals, the vomeronasal sits below the nasal cavity in the palate. In the course of primate evolution, the genes controlling development and function of the vomeronasal in the common ancestor of Old World monkeys and apes began to decrease (Liman and Innan 2003) in favor of color vision (Winter et al. 2003). While it may still function, even in humans, what it does and how it does it are not clear. In nonhuman mammals, the vomeronasal picks up subtle pheromone odors and is acutely involved in sexual behavior (Kimchi et al. 2007). When a cat drops its lower jaw and inhales-exhales over a scent, the vomeronasal is activated. Lack of acute color vision may have been an asset to primates in the deep forest where so many colors fade into blends of black or gray, but in open woodland the ability to perceive a variety of colors and their various shades would have become increasingly adaptive.

Mammals sense the progression of the seasons from the duration of the nighttime rise in melatonin. The *critical day length* (CD) is the minimum day length, or photoperiod, that will induce a physiological response. In most mammals, critical day length is the trigger activating the breeding season. Seasonal breeders depend on the pineal gland to determine the timing of reproduction since the pineal gland secretes melatonin according to the amount of daylight, which in turn depends on the time of year (Moore-Ede and Moline 1985). Some Old World monkeys are seasonal breeders while many are not. Even nonseasonal breeders, however, show a peak of conceptions (or births) over the course of the year. For decades, primatologists had wondered what might be the trigger for reproduction in seasonally reproducing nonhuman primates. Rainfall? Temperature? Sunlight? Moonlight? Clearly, adaptation to energy expenditure in the acts of breeding (conception, gestation, birth, lactation, spermatogenesis, protection, and nurturance) should occur at the optimal time. This means that at least one of these activities should be timed to coincide with food availability (Brockman 2005), especially for the mother. Some primates coordinate reproduction to ensure the best food supply during the energy-expensive period of lactation. Others provide for conception or gestation (Brockman 2005). Any of these factors might be triggered by a local proximal cause, like temperature, but light is a primary factor for most animals, possibly because light is a function of the earth's rotation, while temperature is affected by a variety of more local factors including rainfall, cloud cover, geographic position (like mountains or valleys), or airborne particles from volcanic eruption. Hence, temperature would be a less efficient entrainer *per se*, although it corresponds to seasons that ultimately come from the rotation of the earth and therefore the sun.

In prosimians and monkeys (both Old and New World), there is seasonal variation in birth distributions that can be explained by a combination of what they eat, how big they are (body mass), and where they live: the higher the latitude, the farther from the equator, the greater the seasonality in food abundance. Which aspect of reproduction is in concert with seasonality of food depends very much on the primate type. As noted, even year-round, breeding Old World monkeys and apes experience peaks in part of their cycle. In chimpanzees, for example, recent studies in a variety of locations, like the Kibale Forest in Uganda, indicate that conceptions, rather than birth or lactation, seem to correspond to increased food supply (Sherry 2002).

Reproduction in plants is triggered by variations in light close to or sometime just after the autumn and spring equinoxes. At the equator, however, while they are a great deal more subtle, variations do exist and plants themselves synchronize their flowering and bud formation to a photoperiod (Borchert et al. 2005). Synchronicity is present when, after a time of no flowering, 75 percent of observations of flowering occur during the same period of the year but in different years. No other cue, not even rainfall or temperature, can explain the identical sequence of flowering times as adequately as does the power of light. Recent studies show that even at 1 degree away from the equator, there is sufficient variation in sunrise and sunset to trigger reproductive events. Of these two factors, the light at sunset seems to be the critical one (Borchert et al. 2005).

Plants, like animals, use pigment systems that are sensitive to changes in light intensity. These plant pigments measure changes in the length of the solar day and, as in animals, entrain their endogenous circadian clock to the light. This is new information, and the mechanism controlling plants is not yet well understood, although experimentation has confirmed the phenomenon. Ordinary rice, for example, which is grown at $2°N$ under experimental conditions, reacts to a minimum change in day length of only 15 minutes. This small change is sufficient to trigger flowering. Indeed, both serotonin and melatonin are found in plants (Murch and Campbell 2001), indicating mechanisms similar to those of animals. Because animals eat plants, the hormones ingested, especially the phytoestrogens, will affect them (Parent et al. 2003). These are compounds in plants that have been identified as impacting even humans. The best known of these are the isoflavones, found especially in soy, and which have become popular of late to alleviate the effects of menopause. Over three hundred plants are cited as containing varying levels of phytoestrogens (Mazur 1998). Most of the plants tested are domesticated plants like wheat and peanuts; hence, it is possible that their impact on reproduction would have only been on humans—the primates that invented agriculture. On the other hand, the groups to which these cultivars belong have wild members: grains and cereal plants; oilseeds and nuts (e.g., clover seed); berries and currants; fruits; legumes (e.g., peas); and the allium group, which includes onions and garlic, and so it is possible that hominins ate them long before farming was invented. Researchers suggest that nutrition plays a major role in mammalian reproduction (Cameron 1996; Bogin 1998), providing basic nutrients, energy, and hormones like melatonin. Food plays an important role in determining fertility, fecundity, and seasonal responsiveness. Food restriction can cause the cessation of menses—amenorrhea—in women when body fat falls below a certain threshold, as Ruth Frisch first discovered in the 1970s among ballet dancers and athletes (Frisch 1978; Frisch 2002).

Hormonal Triggers of the Reproductive Sequence in Human and Nonhuman Primates

Recently, the gastric hormones *leptin* and *ghrelin* have been viewed as possible triggers of the reproductive sequence in seasonal-breeding, nonhuman primates. Their activation would explain the endogenous surge that leads to reproduction. Since the beginning of this century, studies on leptin and its antagonist, ghrelin, have given new insights into hunger and appetite (Mustonen et al. 2001; Steiger 2004). These hormones are much in the news nowadays as possible means of decreasing the epidemic of obesity. Ghrelin is active in promoting the organism to eat and therefore causes an increase in body mass and the conservation of body fat (Cummings et al. 2005). It also stimulates the secretion of growth hormone. Leptin, as its "opponent," is the satiation hormone and signals sufficiency. Leptin is secreted by fat cells as well as the pituitary gland in the brain. It informs the brain of how much fat tissue the organism has and therefore has a major role in energy balance, indicating, in the case of reproduction, that the organism has sufficient stored energy to undertake this expensive task. Paradoxically, the leaner the person, the more apt she or he is to have sufficient leptin; the more adipose tissue, the less active is leptin. Ghrelin and leptin are *permissive* rather than *determinative*. That means they may be necessary but not sufficient to activate reproduction. Malnutrition delays the onset of sexual maturation and affects sexual behavior. This is because the cascade of hormones affecting the gonads does not flow, probably because there is less secretion of gonadotropin-releasing hormone (GnRH), which controls the secretion of hormones affecting reproduction (Hileman 2000). If melatonin is suppressed, GnRH can work. Leptin also works to clue the body in to reproductive status. Melatonin, however, regulates the secretion of leptin (Mustonen 2001) and circulates environmental cues. A fall in the level of leptin impairs fertility as it sends a signal that the organism does not have sufficient energy for reproduction (Hileman 2000). Melatonin also interacts with insulin, thereby increasing the expression of leptin that is stimulated by insulin (Alonso-Vale et al. 2005).

This hormone is the *key* as far as reproductive behavior is concerned, as it is fundamental to reproductive maturation. Gonadotropin-releasing hormone (GnRH) is produced by nerve cells specialized for that task, which intermittently send pulses of the hormone out from its source in the hypothalamus to the pituitary gland. There it stimulates the manufacture and secretion of reproductive hormones collectively known as gonadotropins. These enter the bloodstream to reach their target organs, the gonads, which make sperm and mature ovarian follicles. Within the brain, however, GnRH reception is vital to the development of sexual behaviors, from foreplay to consummation.

Curiously, the number of neurons that make GnRH is quite few, relatively speaking—just one thousand to three thousand. And even more curiously, although they are sparsely distributed, these neurons have the ability to communicate with each other and coordinate their secretion activities. Their embryonic history is unexpected, too. They originate in one area, but then migrate to take up residence in several parts of the brain (Sisk 2004).

GnRH is active in most mammals in late prenatal and early postnatal life, helping to spur development into different genders. It becomes quiescent thereafter, but is reactivated at puberty, apparently triggered by developmental processes including body weight. Then it determines pubescence.

Puberty occurs in relation to energy balance; when growth reaches a certain point, there is renewed gene action. The growth information starts a cascade of events. Puberty is biologically "costly" and, hence, requires adequate stores of energy. Puberty is a physical and mental process regulated by neurohormones. The process begins in the brain with the hypothalamus secreting pulses of GnRH. This stimulates cells in the pituitary to release luteinizing hormone and follicle stimulating hormone, which in turn activate ovaries and testes to produce male and female hormones. It is these latter that produce the visible manifestations of puberty one to two years later. A secondary source of secretions determining puberty is found in the adrenal glands. Adrenal androgens, which are independent of pituitary action, stimulate growth of pubic hair, changes in body odors, and skin secretions. Human female puberty has several distinct aspects. It begins with breast development in the earliest teens, which usually precedes menarche, the time of first menstruation, by about 2 years. The age range for the onset of menses in modern humans is between 8 and 13 years, depending on geographic location and socioeconomic class (nutritive state) (Parent et al. 2003). In males, testicular development is the first sign of puberty, usually beginning at around age 11. Melatonin is clearly implicated in puberty, acting initially to incite puberty and then later when puberty is underway (Parent et al. 2003).

As of 2005, evidence suggests that it is the *decline* in melatonin as the adolescent matures that causes puberty (Brezezinski 1997). Increased exposure to light, from staying up later in an illuminated environment, may be the critical factor in stepping up the onset of puberty, regardless of whether the light comes from natural sources or elsewhere. Although some scholars (Goldman 2001; Wehr 2001; Roennenberg 2003) suggest that the Industrial Revolution began the first significant exposure of humans to more constant light, I think it may simply have reinforced and accelerated an already existing condition. The original impetus, I would suggest, occurred much earlier, since we know that both firelight and industrial light suppress melatonin.

In reconstructing the effect firelight may have had on the reproductive system in early hominins, the chimpanzee provides the best model, with the caveat that apes, too, have undergone millions of years of adaptive evolution, and what they are today may not be what they were millions of years ago. Nonetheless, interruption of "normal" darkness would have consequences for all systems, whether of humans or nonhuman primates, affected by melatonin. The suprachiasmatic nucleus and the *pars tuberalis* of the pituitary gland of the brain have long been known to be melatonin receptors. Melatonin directly affects female levels of estrogen and male daily testosterone surges (Buijs et al. 2003) and is ultimately responsible for the onset of puberty. More recently a variety of reproductive tissues have also been identified as receptors. These include male tissues—testis, epididymis, vas deferens, and prostate—and female tissues, including the ovary and mammary gland. Melatonin's actions on these sites are summative; that is, cumulative and increasing. The result is powerful, albeit not always direct, photoperiodic control over reproduction (Shiu 1998).

Melatonin goes through neuronal connections to the autonomic nervous system (which controls, for example, breathing, the beating of the heart, and other vital processes) and provides the anatomical basis for circadian control of the pineal and adrenal glands, pancreas, liver, ovaries, and other organs. Production of melatonin is suppressed when exposed to bright white-yellow light, and particularly, blue light at about 446–77 nm (Lockley et al. 2003). Sleep patterns, in particular slow-wave activity and rapid-eye-movement phases, are profoundly affected by blue light (Munch et al. 2005), as are cognitive processes (Vandewalle et al. 2007). It is this wavelength that triggers a change in the timing of a cascade of hormonal events.

Melatonin has been in the press a good deal this past decade as its role, especially in regard to sleep, has become increasingly understood. It is suggested as a palliative for jet lag and has been proven to be helpful for some types of insomnia and depression, especially seasonal affective disorder (SAD). It is now known that one of the major functions of sleep is to provide a time for the production and circulation of melatonin, which also functions in the body as an important antioxidant (Siegel 2005). Humans, incidentally, require more sleep than other omnivores and less

Table 3. Illuminance and lux values

Illuminance	Lux
Sunlight alone (maximum)	102,000
Skylight alone (maximum)	16,000
Dull day	1,000
Dusk/twilight	10
Campfire	10
Deep twilight	1
Moonlight	.1–.4
Quarter Moon	.01
Starlight; Moonless clear night sky	0.001
Moonless overcast night sky	0.0001

Sources: http://webvision.med.utah.edu;
http://lighttherapyproducts.com/sadinfo.html;
http://www.themeter.net/electromagnetiche.html.

than carnivores, but are otherwise not distinguished in sleep patterns from other mammals (Siegel 2005). Sleep is a time for learning (Hobson 2005), although patterns of sleep vary enormously between human groups (Worthman and Melby 2002). Although melatonin is present at low levels during the day (Brainard et al. 2001), melatonin peaks in darkness two hours after nightfall (Goldman 2001). Entrainment, the coupling of a stimulus with the circadian rhythm, occurs during the transition from dusk to dawn. The mesopic light at dawn and dusk reaches about 10 lux.

The marvel and complexity of this system has melatonin, produced in the pineal gland from stimuli sent by the suprachiasmatic nucleus, return to the suprachiasmatic nucleus and affect neuronal activities there, the products of which in turn cascade throughout the body. The innate rhythm is thus synchronized—that is, *entrained*—by environmental light to the 24-hour solar day, bringing the organism into harmony with the astronomical period (Sancar 2004). Hormones, like melatonin, manufactured in the brain and found throughout the animal kingdom, are the flexible means by which this is achieved. The unity of all living things lies in this fundamental accord. The biological clock is a complex network that has become increasingly complex over evolutionary time. There are multiple receptors and input pathways that feed back and interconnect. Changes in the system, if not acted upon by some other agency such as natural selection, are considered transient, and the clock system will revert to some former phase cycle (Roennenberg et al. 2003). This is important in two ways: first, in considering seasonality in breeding, for example, where experimentation has shown that animals entrained to a particular breeding schedule will revert to the intrinsic schedule once the experimental lighting conditions have been removed, and secondly, in recognizing that a phase shift can be acted upon by natural selection (Roenneberg 2004).

Issues in the Emission versus Reception of Light

The emission of light and the reception of light are quite different phenomena. The light leaving a source—luminous intensity—and the light reaching a target—illuminance—can be measured, but that quantity may not actually reach the retina, complicating evaluation of its impact (Lam 1994; Rea 2002). This discrepancy happens because where the gaze falls in relation to the light source affects the quantity and quality of light reaching the photoreceptors. The direction of the gaze has serious implications for the importance of firelight in this investigation since it raises the question of whether the Ancestor had to have been gazing in the direction of the campfire in order for the light to "register" in the brain and even how much light would enter the system if the person were lying down. A recent study clearly indicates that light coming into the retina from above, so that it hits the lower part of the retina (the system works upside down), is more effective in suppressing melatonin than is light coming from below. The inferior retina is therefore the more important contributor to light-induced suppression of melatonin at certain photon dosages (i.e., light intensities). This suggests that the photoreceptors of the inferior retina are either more sensitive to light or more densely distributed (Glickman et al. 2003). The importance of this finding, to me, is that it indicates that people could lie in a variety of positions near a campfire and still be susceptible to the effects of the light.

But would a campfire have produced the light necessary to have had a physiological effect on the Ancestor? That question is at the crux of this book. In over two years of research, I was unable to find a publication detailing the spectrum of light produced by a campfire, so I decided to conduct an experiment to find out and enlisted friends and students to help. Campfire is incandescent light; that is, it is light produced from heat, as is the light of the sun. Recall the Bunsen burner, or the simple candle. At the base burns the hottest, blue flame (Horack 1997; Kunzig 2001). The coolest, white-yellow flame is toward the top—which is why you can run your finger horizontally through the top of the flame and not get hurt. For both the Bunsen burner and the candle, the fuel comes from beneath, whether as wax or gas (Kunzig 2001).

In order to assess the fire we built outdoors on a June night, we had a photometer sensor that measures light in lux. But before we could assess the fire, we had to build it. First we cleared the ground, then piled up the leaves, stalks, twigs, and branches as we thought perhaps the Ancestor might have done at some point in the millions of years before the genus *Homo* appeared. Our source of ignition was a match rather than a lightning-struck branch or another natural source, but we fed the fire in a casual way, much as people who were not yet habituated to its manufacture might have done. We photographed the fire at different stages and saw the range of colors that characterize the visible spectrum (figure 4).

We were particularly interested in the blue light at the bottom of the fire, which has the greatest effect on melatonin (Brainard et al. 2001). Research since the turn of the twenty-first century has suggested, and now confirms, that blue light at 446–77 nm "turns off" melatonin production, at which point the organism returns to daytime behavior. It was only in 2003 that scholars tested a group of young men and women and found that, contrary to what was commonly believed, it was not the green wavelength at about 555 nm that shifted circadian rhythms, but rather the blue light (Lockley et al. 2003). They found that green light could be effective but required 6.5 hours of exposure and only readjusted the body's clock by 1.5 hours. Blue light, however, after the same number of hours of exposure, had an effect twice that of green light, readjusting the body clock by 3 hours. In addition, blue light, even in daylight, affects cognitive processes. In recent experimental work, nonvisual photoreceptors,

FIGURE 4. The campfire (photo by author).

mediated by melanopsin in blue light, *enhanced* memory tests, whereas green light did not. Studies on how short-term memory becomes long-term memory indicate an epigenetic route (Colvis et al. 2005) and suggest that this remembering of behavioral innovations would affect subsequent behavior in evolutionarily significant ways. Interestingly, the brain responses clearly affecting the circadian system were also found in areas associated with what are called executive functions (where vital monitoring of responses, planning, and decision making occurs) and mostly in the left, verbal hemisphere (Vandewalle et al. 2007).

In our experiment, distance from the fire itself made all the difference in how much lux arrived at the eye. At 2 m (approximately 6.5 ft) from the fire, we measured only 3 lux, but at less than 0.5 m (approximately 1.5 ft) the measurement was up to 50 lux. Cooking distance registered 20 lux. The critical value, however, seemed to be a measurement at eye level taken at just over 1 m (approximately 3 ft) from the fire. With the gaze fixed on the blue light at the bottom of the fire, the lux value ranged from 7 to 10, about the same as at twilight.

Far less light is needed to suppress melatonin than was originally thought (Aoki et al. 1998). Chronobiologists have found that for some men under experimental conditions, full suppression of melatonin can be achieved with exposure to light of about 17 lux, but that others respond readily to only 5 lux (Lockley et al. 2003), and that even 1.3 lux of blue light can measurably affect melatonin levels (Raloff 1998). Significantly, it has now been found that entrainment occurs even *in utero* (Reppert et al. 1985). At least in rats, melatonin has been found to cross through the placenta and has also been found in mother's milk (Vanecek 1999). So maternal transfer of the effects of light does affect the newborn—at least in mice (Aton et al. 2004)—and in monkeys, where the mother's melatonin during the fetal period is a *Zeitgeber* to the growing fetus's suprachiasmatic nucleus, predisposing its circadian rhythms (Torres-Farfan et al. 2006) to sensitization by light. Could it be, then, that mothers' proximity to campfire might have enhanced changes in circadian rhythms in the Ancestor? At this point, it is important to distinguish between epigenetic *effects* and epigenetic *inheritance*; the former affecting the phenotype of the offspring, the latter affecting its genotype as well (Youngson and Whitelaw 2008). Epigenetic effects, however, are known to continue "unto the generations." Changes in the mother's environment—diet or pesticides, for example—can and do affect offspring prenatally, and these are carried forward as the environment in which genes interact (Bjorklund 2006; Heijmansa et al. 2008).

Conclusions from the Campfire Study

Given that the mesopic light at dawn and twilight, which is equivalent in lux to that of the campfire, is sufficient to suppress melatonin production, we concluded that campfire would be able to do the same. There are several issues with this conclusion, however. First of all, our fire and experiment were not laboratory controlled. Secondly, we trained our gaze toward the fire, whereas the Ancestor might have looked around, to the sides as well as across the fire, or perhaps not in its direction at all. Thirdly, it is not known how much exposure is necessary to have profound physiological effects on hormonal balances and whether exposure can be cumulative over a period of years.

Sources of information come from endocrinology, its sister discipline behavioral endocrinology, and socioendocrinology. Behavioral endocrinology studies the effects of hormones on behavior, while socioendocrinology focuses on the reciprocal roles that hormones play on social behavior in mammals and that social behavior plays on hormones. Physiology and anatomy, as well as neuroanatomy, neurology, and neuroendocrinology are also concerned with these issues. Physiological and behavioral rhythms can vary a great deal over the course of a day or a year. A recent study, for example, tells us that if you must go to the dentist and are going to ask for an anesthetic, be sure to go in the late afternoon because the anesthetic will be far more effective (Gorman and Lee 2002). Even chemotherapy's effectiveness depends on the time of administration, and the evidence for such oscillations throughout our daily lives is growing.

The coordination between the organism and light cycles—entrainment—is incredibly flexible. The internal organic system is cued into a different state by the environmental one. Information from the environment is mediated through a visual pathway into organs housed in the brain. Organisms are born with the inclination to be entrained. It is a genetic propensity; the activation of this inclination occurs during development.

Genes exist in an environment of other genes, as well as a variety of substances, and receive information through many kinds of systems from the outside environment. Epigenetics studies this relationship, where transmission of information is not encoded in DNA sequences, but rather by "markers" on the chromosome that activate or deactivate genes and that are then transmitted from cell to daughter cell on down through the generations (Fazzari and Greally 2004). Melatonin may be implicated in this activity, according to a recent study, by affecting nuclear receptors that in turn alter the structure of DNA (Irmak et al. 2005). Nutrition has an epigenetic effect: genes can be the same in two members of a species, but life-history effects, especially nutrition, can make these two look, behave, or cycle entirely differently (Callinan 2006). In this way epigenetics resolves a paradox.

The *epigenome* is a biochemical system lying alongside DNA that operates on genes. Because DNA is so long, it cannot fit into a cell unless it is wrapped and coiled around a group of proteins called *histones*. A "package" of histones and DNA is called a *nucleosome*. The nucleosome is the place where various chemical modifications "mark" the DNA complex and affect the expression of the corresponding genes at that location. The epigenome therefore unites the "interior" of an organism with the environment outside of it. Mother's nutrition affects the growing fetus in utero; nutrition from infancy to old age affects genetic mechanisms. In the process of development, cells that are initially undifferentiated take on different characteristics. For example, liver cells differ from eye cells, but the DNA remains constant. This occurs through a process of "gene silencing," one type of which is where a particular molecule, a methyl group (composed of one carbon atom and three hydrogen atoms), attaches to a nucleosome around which DNA is wrapped. This *methylation* blocks access to the gene, which results in "silencing" that gene, that is, inhibiting its expression—not altering it (Colvis et al. 2005). This mechanism allows for plasticity and responsiveness to changing conditions. For example, levels of folates and B12, among others, affect the levels of a substance known as SAM (S-adenosylmethionine), which "donates" methyl groups providing the raw material for epigenesis. Indeed, as Waterland and Jirtle recently suggested, an overabundance of a nutrient, such as vitamin B12 or folic acid, during pregnancy could even have deleterious effects. In their classic experiment with agouti mice (especially bred for purity of genetic strain), supplementing the diet of the agouti mother changed the coat color in the offspring (Waterland 2003). They concluded that early nutrition has an epigenetic influence in the early embryo and that its effect *continues* down the generations.

There is increasing evidence that paternal influence via sperm confers epigenetic changes as well. Under experimentation, a female rat's endocrine system was chemically disrupted. The effects were transmitted through decreased sperm production in the *male* line for several generations (Anway et al. 2005). Hence, genes without environmental information are insufficient to provide an explanation of physiological or behavioral phenomena. The basic mechanism of the molecular clocks is a gene-protein-gene feedback loop. The protein returns information back to the genes, which determine further manufacture while stimulating the transcription of other clock proteins (Buijs et al. 2003).

Phase shift, as the reader will recall, is the change in onset of occurrence of a rhythm due to a change in its stimulus. The stimulus, or *Zeitgeber*, that we are considering in this book is the light-dark solar cycle. If the phase of the *Zeitgeber* suddenly changes, the organism's cycle will resynchronize with the new phase. This means that *exposure* to campfire light, especially the blue wavelengths, was active in phase shifting the circadian rhythm in the Ancestor and with it all cycles that are hormonally intertwined. Phase-shifted cycles can shift back, but if the *Zeitgeber* influence remains present, that return shift will not be able to take place. The possibility of effects in utero will be important in this change. Interestingly, different rhythms in the body entrain at different rates, thereby affecting the harmony in the organism. Perhaps the lack of quiescence—harmony—is itself a stimulus in human evolution.

The hypothesis that I am putting forward in this book is that the light of the campfire could cause a phase shift, increasing the body's daytime by decreasing melatonin. The anthropological literature has always noted that campfires would "extend the day," but the physiological implications of this were not explored, though the idea was nascent in Pfeiffer's (1971) suggestion that firelight might affect rhodopsin in the eye and consequently estrus. The repercussions throughout hormonal systems and patterns of brain activity over time may be a critical factor in explaining the divergence of our species from kindred apes.

FOUR

Elementary, My Dear *Orrorin*

When the Ancestor First Used Fire

In thinking about the whole process of hominization and the role fire played in it, I have begun to see the hominins as being responsible for the first steps in the domestication of fire. The significance of these first steps lies in their ultimate impact on the course of human evolution itself. I suggest that the more that hominins *did* something to the environment by way of intervening with the natural process—in this case, artificially increasing the amount of exposure to light—the more these hominins dampened the effect of the environment on themselves, and the more they took over the direction of their own evolution. The acquisition of fire is precisely that kind of intervention. But who was responsible? And when did they live?

The current definition of "hominin" takes bipedality as its first criterion, and we know that there were hominins walking about on two legs at 6 mya, as there seems to have been since the Middle Miocene, somewhere around 20 mya (Filler 2007; Maclatchy 2004). We know as well that the divergence between chimpanzees and humans seems to have been achieved somewhere around 5 to 6 mya (Schmidt et al. 2003; Patterson et al. 2006). So why not simply place the chronicle back there, at around 6 mya, toward the end of the Miocene? Since evidence for the *control* of fire does not exist prior to around 1.6 mya (at Koobi Fora), the idea of placing first *use* of fire at around 6 mya has been a totally heretical notion. But I will argue in this chapter that important, new research now supports such a possibility. One controversial study accepts the likelihood of hybridization between the human line and chimpanzee line, which must have diverged from other ape forms somewhere around 10 mya. The two lines continued to interbreed and create hybrids until just around 6 million years ago, when they finally and irrevocably ended up in two different species (Patterson et al. 2006). This work affirms the time period in which the first stages of fire acquisition could have occurred, that is, in bipedal hominins beginning somewhere around 6 mya.

Another objection to placing the first use of fire at about 6 mya is that evidence for the use of stone tools dates to only as recently as 2.6 mya, in Gona, Ethiopia; so why not assume that domestication of fire had its origin then, at 2.6 mya, too? But I don't think so—I think that's too late. One of the problems is that tools made of materials other than stone are hard to find—they rot. However, we know that tools must have existed long before 2.6 mya. Monkeys will use sticks in threat and even as adjuncts to getting food. Apes are well-known tool users, working with sophisticated tools to get difficult things like honey, or even in extravagant displays. Both chimps and gorillas have been seen brandishing sticks or branches during their social demonstrations (Goodall 1986; Wittinger et al. 2007). Fashioning tools from stone was a major undertaking, and the physiological, psychological, and technological changes that were necessary precursors to this endeavor would have taken hundreds of generations. We know that conservatism marks human evolution; the structure of stone tools called Mode 2 did not change for over a million years. So I am *speculating* that the history of fire in human evolution may also have been a slow process and that the time to focus on is an earlier period.

Could it then have taken 4 million years for the effects of fire to become significant? Possibly. Evolution moves both slowly or quickly depending on the circumstances, so evolutionary theory does not impose constraints of time. If the process of the use and impact of fire by and upon hominins began in this 4-million-year time frame, this would constitute

200,000 generations, given 20 years to a generation. And at 15 years from birth to reproduction, that is over 267,000 generations. Clearly, there is enough time for massive change.

I have suggested that there are stages in the domestication of fire. In this book, I am suggesting that perhaps the *stages* in the domestication of fire coincide with the three major periods of hominization. And although these stages toward the domestication of fire and the concomitant domestication of ourselves would remain the same no matter when they occurred, I will argue that knowing just when to position these stages becomes important if, as I am speculating, association with fire actually *caused* humanity itself.

Let us examine, for a moment, what these periods of hominization and these stages of fire acquisition comprise. The *earliest* (as far as we now know) is when the ape and human lines diverged, between 5 and 10 mya (depending on the research). The *middle* period, around 3 to 4 mya, is when the truly human pattern was established, with important changes in bipedalism and probably dietary habits, characterized mainly by the australopiths, followed by a very rapid human change around 2.6 to 2 mya when *Homo* is present. The *final* period, relatively recent in historical terms, dates from around 200 kya and is when our modern species emerged. Recently, fossils from Omo I and II in Kenya were confirmed at dates from 195,000 years ago, and they are now the earliest truly *Homo sapiens sapiens* (McDougall et al. 2005).

Perhaps the *stages* of domestication of fire correspond, then, to these three periods of hominization. First approaching, associating, and finally nurturing fire would correspond to the time of earliest bipedality and thereafter. Control would correspond to the middle period, and full domestication—that is, manufacture of fire—would correspond to the final period. If this were true, the *effects* of fire would have had a major impact at the *first* stage, when intimacy with fire's properties was new, but the transformation would have continued through the second and third stages. That paradigm fits the known facts, at least at the time of the writing of this book. I leave it to you to decide whether this scenario "works."

In the rest of this chapter, we will consider both genetic and environmental evidence for the placement of the chimp-human divergence at about 6 mya and what factors might have prompted the initial and continuing approach to fire by the Ancestor. In addition, the specifics of the course of primate evolution, the importance of bipedalism to the first use of fire, and fossil evidence for the proposed trajectory on the course toward modern humans will be will be presented, and a discussion of what all these investigative paths contribute to reconstructing the Ancestor—the Ancestor who first used fire.

Tracing the Genetic Divergence between Chimpanzees and Humans

Since the revolution in genetic research beginning in the 1950s, our knowledge of genes, gene action, and genic evolution has grown enormously and has enabled scientists to pinpoint the genetic differences between two species and the time it takes for this difference to become established in the population. With the advent of the Human Genome Project and the identification of all the genes in our species, the speed of knowledge in this area has accelerated exponentially. The idea of this project, begun in 2002 and completed 3 years later, was to "map" the entire human genome and to describe the common patterns in human genetic variation. Now that the chimpanzee genome has also been mapped, comparisons between the two genera can be exact, and inferences about gene action, and about evolution, are more secure (Gunter and Dhand 2005). We are no longer talking simply metaphorically about beads on a string but about genes on a chromosome.

Genes interact—a phenomenon known as *epistasis*. Thus, the *genome* is an interacting system of sets of nucleic acids (the building blocks), which are at the core of a chromosome. These groups, or *codons*, determine the sequence of amino acids in a protein. Change, that is mutation, can take place within one of the building blocks (a point mutation), or in the sequence of building blocks on a chromosome that determine the sequence of amino acids. Point mutations occur in about one of every 100 million DNA sites in each generation (Fondon and Garner 2004). Bits of chromosomes can be added, deleted, become silent, or duplicated. These "bits" are sequences of codons; hence, any change in the nucleic acid building blocks or change in the structure of the chromosome affects the end product.

Point mutations may be the quintessential way a gene changes, but larger changes occur when the lineup of genes alters through a variety of chromosomal actions. Parts can be deleted, duplicated, reversed, or turned upside down, and the way their codons read will make the end product very different indeed. In addition, one gene can have multiple effects (Ledford 2008). Temperature or other environmental stimuli can enhance this pleiotropic ability. Point mutations are thought to be relatively slow in their evolutionary impact. So how to account for rapid change? "Punctuated equilibrium" has been the model for evolution for some time. Periods of slow change are suddenly interrupted by spates of rapid diversification. How does this rapid diversification come about? The answer is related to another question: how can the genomes of two species—say, chimps and humans—be so similar and yet their forms (and functions) remain so different? Although 99 percent of the genetic material is the same between chimps and humans, with only 1.2 percent of the nucleotides (containing the "letters" of the genetic code) being dissimilar, we now know that the rapid diversification between these two species has happened largely with respect to this tiny percentage of the genetic material known as the "developmental genes," those genes responsive to and responsible for development.

The particular sequences that are liable to mutation are repeating sequences of the chromosome known as *tandem repeat sequences*—these are responsible for rapid morphological change. A mistake occurs in how many copies of a particular sequence are made during the copying of DNA. For example, a point mutation in a gene codon "cat-cat-cat" might read "act-cat-cat," but a mistake in a tandem repeat sequence might read "cat-cat-cat-cat-cat." The resulting production from this segment will be quite different. Mutations in tandem repeat sequences occur more often than do point mutations—and much more rapidly, almost 100,000 times more rapidly—and have a rapid and powerful effect on the physical appearance of an organism.

As an example, Fondon and Garner, scientists at the University of Texas Southwestern Medical Center in Dallas, studied ninety-two breeds of dogs, of which the variety of forms is truly impressive. Genes involved with development of limbs, snout shape, and size were the particular focus. They found evidence for selection of divergence in tandem repeat sequences relating to the development and the ultimate form of the face of the breeds. Variations in the number of repeats in a tandem repeat sequence were distinctly associated with significant differences in limb and skull morphology, for example, short or long muzzles (Fondon and Garner 2004). Comparing domesticated dogs to wolves and coyotes showed, as you might expect, that domestication was associated with more variation in the lengths of repeated sequences in the domesticated dogs than in the wild forms. Domestication of animals tends to be an identifiable process (Zeuner 1963) wherein certain phenotypic changes are typical: foreshortening of the muzzle, blanching of the fur, crowding of the teeth, and even changes in the amygdala in the brain relating to fear (Hare and Tomasello 2005). In general, a relaxation of natural selection occurs as nonadaptive traits are supported by human care, a process seen in sheep and mice, as well as dogs, and whatever other animal has been domesticated. So stability of form is apparently relaxed under domestication. The implication of this for human evolution is clear. The acquisition of fire is precisely the kind of domesticating influence that could have resulted in greater genetic variability among our hominin ancestors.

There is other evidence supporting the possibility that tandem repeat sequence evolution might be a mechanism underlying hominin diversity. A recent study by Ackermann and Cheverud (2004) examined whether forces other than natural selection have been at play in human evolution. They had in mind the action of *genetic drift*. Genetic drift is a random process altering the frequency of variants of a gene from one generation to another. It played an important role in hominin evolution, especially, according to these authors, at around 2.7 mya, at the time of the diversification of species in the genus *Homo*. There are mathematical formulae that can calculate whether genetic drift rather than selective forces has been a factor. Applying these to a sample of seven genera of hominins led Ackermann and Cheverud to conclude that while some traits were clearly adaptive in earlier genera, especially *Australopithecus*, other traits in *Homo* seemed to be *randomly* established. This was the first time such a study was applied to early hominins. Their approach focused on variation across physical traits, emphasizing the *pattern* of variation rather than the *total* variation. For example, australopith features such as big bones and crests on the skull suggest specific adaptations, but the kind of variation seen in *Homo*, especially the random pattern of variation in facial features, indicates other processes at work.

Another major mechanism for the divergence of apes and humans is through *gene duplication*. Since the Human Genome Project, the precise chromosomes and the genes on them have been documented, and the number of gene differences counted. Chromosome 5 in humans, for example, derives from chromosome 4 in the apes; chromosome 21 in humans is equivalent to chromosome 22 in apes. The chromosome 21–22 link has been the subject of exhaustive investigation, and differences in this single chromosome and the genes it carries account for a substantial number of differences in protein end products controlled by those genes.

The differences between our two groups are due to *indels*, that is, insertions and deletions along the chromosome. These are the basis for protein diversity. As if all of this were not complex enough, a single gene can have multiple transcriptions; it can cause the manufacture of more than one end product, explaining the rapid differentiation between the higher apes and humans (Consortium 2004). The differences have been dated, and the function of duplicated or lost genes established. The ratio of increased to lost genes in humans is 134 to 6 (Fortna et al. 2004).

Other studies show profound differences in certain genes: twenty-seven of forty-eight olfactory proteins showed accelerated change in humans, whereas chimps showed increases in gene proteins related to body structure, especially muscle mass (Clark et al. 2003). The "gracialization," that is, the development of gracile or delicate-bodied humans (compared to chimps) owes much to the deletion of certain genes or part of the gene complex. Geneticists have been able to locate and date the exact change that took place in a gene, *MYH16*, involved in muscle development. The point mutation occurred about 5.3 mya, according to the genetic researchers (Perry et al. 2005), within the time period of the divergence between chimp and human.

Direction of Genetic Change Supports Selection for Cognitive Development in Human Evolution

Primates, including humans, differ from other mammals in the nature of their brain and the capabilities that structure permits. Human evolution is characterized by increasing reliance on the brain over the physical structure of the body. This reliance promotes a positive feedback system wherein the innovations designed by the brain in turn affect anatomical structure and physiology. In other words, brain development ultimately fosters further brain development. Genetic research continues to support this picture, as we will explore below.

Recent studies indicate, for example, that a particular gene, a member of the family *HAR* (human accelerated regions), seems to be at the root of the relatively fast changes that took place and that differentiate the brains of chimps and humans. It is part of an RNA gene that functions to influence developmental process by acting on the migration of neurons in the neocortex (Dorus et al. 2004). The authors of this study knew that the regions they were looking at were conservative in mammalian animals but had undergone rapid change since the last common ancestor. One of these genes has, in a relatively short time span, accumulated eighteen changes in the sequence of nucleotides (Pollard et al. 2006). This many changes are beyond chance—it distinctly suggests selection for cognitive development.

Further significant genetic change between our Ancestor and non-human primates has been found in a molecule that operates primarily in the brain, retina, and testis. The gene controlling this material is itself descended from a gene that insinuated itself into the genome and was then perpetuated. It did this through the process of reverse transcription, where messenger RNA transcribes to DNA, instead of DNA to mRNA, the latter being the route once considered "central dogma" in the field of genetics (Brosius 2003). The transformation of this molecule took place sometime after 23 mya and before 18 mya. It clearly distinguishes among lesser apes (gibbons), apes and humans, and Old World monkeys. More importantly, identifying this genetic change indicates that the selective pressure on neuronal modification and brain growth began as early as 25 to 30 mya (Stewart and Disotell 1998), well within the new dating of the last common ancestor between Old World monkeys and hominids (29–34.5 mya) (Kaessmann and Burki 2004).

The most pronounced increase in *gene number* has occurred in humans since the divergence from chimpanzees. And although we have lost functional genes—for example the ability to manufacture certain acids (sialic, in particular), an ability that chimps have retained—the increase in gene number has been mostly in those responsible for neuronal and brain development. In 2006, for example, a team of geneticists found that a gene that codes for a particular protein is represented by 212 copies in the human, but only 37 in the chimp. This protein is found in many places in the body, including the brain (Popesco et al. 2006).

One of the genes newly identified through comparison of chimp and human genomes controls synaptic branching—the increase in communication between neurons—while another controls an enzyme involved in learning and memory. A recent study has located a gene, called *ASPM*, the function of which is not yet known, but which is important because it shows positive selection for brain development; that is, it makes proteins different from those in other species of mammals, including chimpanzees (Dorus et al. 2004). This *ASPM* gene, like the *HAR* genes mentioned earlier, demonstrates rapid change in the primate lineage leading to modern humans and might explain the size of our cerebral cortex, the site of abstract reasoning that is extremely large in humans. At the lower limit, then, about a hundred genes are today thought to have undergone major changes in activity as it relates to human brain evolution (Preuss et al. 2004), affirming the emphasis on brain function, especially communication to and from this organ and the body. These genetic changes suggest that the trend toward "mind" as the major means of adaptation has an early history.

What strikes me as most exciting about these various lines of genetic research is in (1) finding a mechanism for rapid evolution—we need that mechanism to explain aspects of hominization; (2) confirming that domestication relaxes genetic stability; and (3) noticing that the direction of genetic change between chimps and humans supports the role of increasing cognitive capacity in determining further human evolution.

Environmental Sources of the Chimp-Human Divergence

The separation of chimps from the Ancestor also owes much to the environmental changes that took place in Africa (Maslin and Christensen 2007). The formation of the Great Rift Valley in eastern Africa began as early as 45 to 33 mya, in Ethiopia, and spread to northern Kenya and Tanzania, by between 15 and 8 mya, and by 1.2 mya had produced substantial variations in elevation in Tanzania (Maslin and Christensen 2007), with most of the uplifting taking place between 7 and 2 mya (Gani and Gani 2008), coinciding with major developments in human evolution such as the chimp-human divergence and bipedalism. Today it stretches 5,632 km (3,500 mi), from Mozambique to the Red Sea, and in some places reaches depths of 609.6 m (2,000 ft). The tectonic movements created mountains to the west, where moisture was trapped, and precipitated the creation of an environment that remained moist and wooded, while the eastern side became increasingly dry, slowly becoming savanna and open woodland (Maslin and Christensen 2007). Chimps are principally found today in the Rift Valley or west of the Rift, while human fossils are found in the Rift Valley or to the east (Chad is the exception). Evidence for orogenic (mountain-building) change comes from the earth itself; evidence for climate comes from plant fossils and the tissues of ancient animal forms, particularly bioapatite, the major constituent of tooth enamel. Analysis of the type of carbon (the isotope) in these tissues indicates which plants were available and what the climate might have been.

Over geological time, some animal species have become more abundant while others have declined. There has been some controversy as to what was responsible for the change in this abundance; local climate change, migration, and other factors were considered, but it is now clear from stable carbon-isotope studies that an important global ecological change was occurring in the amount of atmospheric carbon dioxide (CO_2). An asteroid impacted the earth about 8.2 million years ago and apparently caused a cooling spell, which lasted about 1.5 million years (Farley et al. 2006). This lowered the amount of atmospheric CO_2, which favors what are known as C4 plants over C3 types. In addition, the faulting along the Rift Valley increased aridity, favoring C4 plants (Maslin and Christensen 2007). The numbers (3 and 4) refer to the end result of photosynthesis, which fixes CO_2 in a three- or four-carbon compound. Over 95 percent of plants on earth are C3 plants, including, for example, trees and some grasses, while C4 plants include many of the common cultivated crops, like sugar cane and corn, grasses and sedges. In addition, C4 plants include the seeds and storage organs of sedges, grasses, and forbs. C3 plants thrive in CO_2 concentrations above 500 ppm (parts

Table 4. Epochs and their dates

Epoch	Date
Holocene	10,000–today
Pleistocene	1.8–10,000 mya–kya*
Pliocene	5.3–1.8 mya
Miocene	23.8–5.3 mya

Note: *Millions of years ago (mya) and thousands of years ago (kya).

per million), while C4 plants prefer a lower level, which is what was happening by that early date (Cerling et al. 1997). The environmental change had been a long time coming. Grasses, which were so important in creating the environment that fostered human evolution, had been around a lot longer than originally thought, predating the Ancestor by millions of years.

The period spanning the Late Miocene and Early Pliocene, and accompanying the isotope variation, was a time of worldwide faunal change. This was the period giving rise to the Great Rift Valley in eastern Africa. The epochs are listed in table 4, and the date when the period is *traditionally* thought to have begun is given in the second column (in millions of years ago).

Early Miocene mammalian faunas in eastern Africa had a tropical-forest character with common taxa including hominoids, hyraxes, piglike animals, rhinos, and elephantlike animals. Evidence from a variety of sites all over the world, and at important hominin sites in eastern Africa, was taken from the teeth of proboscideans, elephantlike animals, and equids, the horse group. Horselike creatures are known from about 10 mya, and their teeth show a diet of grasses containing C3, hence higher values of atmospheric CO_2. By about 7.5 mya, these animals were ingesting a predominantly C4 diet. The elephantlike animals follow the same pattern, but more recent in time.

Fossil diets are inferred from a variety of sources. These include microwear patterns on teeth, isotope analysis, biomechanics, and, of course, plain anatomy—as in tooth size or length of shearing crests on the molars. Wear patterns, observable with the electron microscope, can be identified by comparison under experimental conditions. Techniques in biomechanics can be used to analyze the stresses and forces on bone and tooth under different dietary regimes (Lucas et al. 2008).The researcher can see what kind of scratch or abrasion a particular pit, bark, or other object makes on the tooth. Isotope analysis biochemically evaluates the manner in which CO_2 is taken up by plants. The proportion of C3 to C4 plant foods in the diet can then be measured.

The leaves of forbs (for example, milkweed, purple coneflower, fireweed) provided some dietary essentials: minerals, vitamins (including, occasionally, vitamin C, which primates cannot manufacture in their bodies), and a good amount of moisture. However, the amount of carbohydrate, fat, and especially protein provided in the leaves of forb plants is severely limited. Hence, the Ancestor would have had to find these nutrients elsewhere—in insects, for example. In the earlier period, the sources of nutrients would have included micromammals (such as dormice), small reptiles, insects and other invertebrates, and bird eggs and nestlings (Peters and Vogel 2005). Later, around 3 mya, the list would have expanded to include small ungulates (like dik-diks, which are smaller than goats) and even larger hoofed mammals. While C4 grasses offer seeds, tubers, bulbs, and rhizomes, the larger underground storage organs, like tubers, would probably have been unavailable to the small-bodied Ancestor, with neither the strength nor the tools to extract them (Peters and Vogel 2005). About 30 percent of the Ancestor's C4 intake would have come from eating organisms that ate C4 plants. Contemporary baboons eat such small mammals, killing them as they encounter them in the grass—but not usually hunting them as do chimpanzees, although hunting traditions in baboons can occur and then fade (Strum 1987). I have seen macaques dig down about 10 cm (3.9 in) to pick out bulbs and rhizomes in the soil, whereas chimps can dig considerably deeper.

Professor Augustin Kanyunyi Basabose, a primatologist at the Centre de Recherche en Science Naturelles in the Democratic Republic of Congo, tells me that chimpanzees in Kahuzi dig down 40–50 cm (15–19.6 in) into the ground searching for food, and they even do it regularly. They dig ground to search for subterranean honey and larvae of

Meliplebeia tanganyikae, which is a "stingless bee nesting underground," as well as tubers, using tools to get at them (McGrew et al. 2005). But tubers that are even deeper (approximately 129 cm, or 48 in), such as those the Australian Aborigines depended on, require the mind and tools of *Homo* not only to dig them out but also to detoxify them, as many are either inedible, toxic, or digestively unacceptable since they constipate or are purgatives (Peters and Vogel 2005). One scholar places the use of these potential foods at about 1.9 mya, suggesting that fire was necessary to make them edible (Wrangham et al. 1999). Seeds are a good source of minerals, vitamins, and carbohydrates and afford some protein. However, gathering seeds is time consuming and converting them to food—like a tortilla or porridge—requires tools to husk and separate chaff from kernel, grinding implements, and a cooking fire. These were not available at so early a point in history. It is worth noting, however, that traditional peoples, like the Aranda of Australia, gathered seed heads and chewed on them as they walked, and that the hard foods and small objects have been calculated through biomechanical analysis to constitute part of the diet (Lucas et al. 2008).

During the Late Miocene (5.3 mya to 11.6 mya), open-wooded, grassland habitats replaced the earlier, less seasonal woodland-forest habitats. Divergence and radiation of the various monkey types took place, and the number and kinds of apelike primates decreased, eventually to be reduced to one Asian and three African types. The forest monkeys may be responsible for the demise of the variety of apes with their habits of eating unripe fruit and chewing up seeds, killing off the new forest, and ape habitat, before it had a chance to grow (Dominy and Duncan 2005). The subsequent epoch, the Pliocene, dating from 1.8 to 5.3 mya, witnessed a sharp increase in seasonality. The animals contemporary to that epoch were adapting to savanna-woodland. Grazing antelopes and hippos became more common, replacing earlier forms, as did three-toed horselike animals and elephants with high rather than low-crowned teeth, which they have today. As will be discussed, this evidence of climate change helps to explain why early hominins and their bipedal adaptation emerged, taking into account that human evolution took place in the context of a "C4 world" between 8 and 6 mya, when grasslands were spreading quickly.

The Course of Primate Evolution

There has been a recent trend to move all dates back as we consider the course of primate evolution. Accordingly, the first primates appeared not 65 mya, as has been presumed for decades, but probably about 85 mya, and the divergence between Old World monkeys and the hominids (apes, gibbons, and humans) most likely occurred somewhere between 23 mya (Raaum 2005) or 29 to 34.5 mya (Tavare et al. 2002). Similarly, the divergence between apes and humans, assumed for a long time to have taken place at about 5 to 6 mya, has now been pushed back perhaps to between 8 and 10 mya (Tavare et al. 2002). With the new, albeit controversial, find between 10 and 10.5 mya of a possible ancestor to gorillas (Suwa et al. 2007), and the slightly younger ape find from Nakali, Kenya, which has been dated to just under 10 mya (Kunimatsu et al. 2007), the divergence within the ape to human group seems to favor the earlier date. The divergence between chimpanzees and humans, however, seems to have taken place somewhere around 5 to 6 mya (Schmidt et al. 2003; Patterson et al. 2006).

That prosimians—the group represented by the tarsier or the dwarf lemur—gave rise to subsequent types of primates is now incontestable. Although the new date when they first appeared has recently been set back 20 million years, the sequence remains familiar (Tavare et al. 2002). It is now thought that the last common ancestor of the primates was similar to a dwarf lemur and lived in tropical or subtropical forests, with a diet composed mainly of fruit and insects. Most nonhuman primates are tropical, with most fossils being found in the tropics as well, although— importantly—at varying altitudes. There are only a few species of primates that live outside the tropics or subtropics, the best known of which are the Japanese macaque (some live in the mountains) and the monkeys of Gibraltar (Mediterranean). All of the apes, New World monkeys, prosimians, and most of the Old World monkeys, except those noted, are tropical, where seasons are marked by changes in rainfall as opposed to temperature.

The last common ancestor of monkeys, apes, and humans was vertically oriented toward the tree trunk rather than horizontally quadrupedal on a branch, foreshadowing a major alteration in focus and relationship with the world. This earliest pre-primate ancestor was ratlike, with eyes almost lateral on its triangular skull, and the opening hole at the base of the skull would have been at the back of the head rather than at the base of it. The five digits of the hand could come together to grasp, but both hands worked together as do the hands of squirrels, with the thumb a mere appendage.

The ancestral primate, not to be confused with the pre-primate creature just discussed, would have been nocturnal, lived in trees, weighed about 450–900 gm (1–2 lbs), and had grasping hands and feet. Its eyes would have faced forward and had overlapping fields of vision, giving it depth perception. Its snout would have been short compared to its predecessors, emphasizing a visual brain over one primarily based on smell. It most probably had offspring one at a time, and, like many contemporary prosimians, she might have "parked" the infant in a nest while she went off to forage, but the dependent offspring could probably cling to its mother with hands and feet. The basic primate pattern was set.

What happened in primate evolution was a total reorientation of the body. As the body came back on its haunches, the shape of the pelvis began to change, with the bony plates on the side slowly, over eons, becoming shorter. The snout shortened, and gradually the sense of smell gave way to improved vision. The eyes came to occupy a front position on the face, and each eye could see over the range of the other—overlapping fields—permitting depth perception. The period encompassing the transformation from the pre-primate to the Ancestor lasted 60 to 70 million years—a lot of changes took place in that time; I have not touched on the teeth, physiology, brain structure, or the myriad of other changes that took place as well.

The most dramatic and significant change of all was in the hand. Even earlier than New World monkeys, prosimians had hands that could be used independently of each other, with (mostly) five mobile digits (not all prosimians have five). New World monkeys cannot really use their thumbs as do Old World monkeys, apes, and humans. While they have tails that can grasp like a hand, and which is complete with "fingerprints," their thumbs do not join the hand as do those of Old World monkeys. Apes, Old World monkeys, and humans can rotate the thumb on its own axis so that the soft part of the thumb faces the soft part of the other fingers. This motion, called opposition, is a hallmark of hominization. Yet monkeys of the Old World (baboons, macaques, mangabeys, etc.) all have this ability and can do it even better than apes. Apes have elongated digits and a shortened thumb. If they want to pick up a small object, they would do so in a scissor grip, using the second and third digits—the thumb is not used. Monkeys have a hand with proportions of digit, palm, and thumb like ours, and so can readily manipulate objects. My doctoral research compared the hand in six species of African forest monkeys with a focus on the muscles of the thumb. There was an interesting curiosity: while the thumb had the *potential* to rotate and face inward—what is termed "opposability"—in truth, it could not do "opposition" because the muscle and tendon required to hold the thumb sufficiently erect to form a tweezer grip was missing (Burton 1972a), so the monkeys had to lean the outside of the thumb against the side of the second digit instead.

In the course of primate evolution, eye-hand coordination was enhanced and selective pressure was placed on cognitive processes in the brain. The homunculus drawing of the brain, so popular in Psych 101, shows the motor cortex of the brain and what percentage of it is devoted to specific motor functions—and a great deal of the motor cortex is taken up by the thumb. It is, in my estimation, the most important organ in primate evolution. Freedom of the hands and opposition of the thumb reoriented the body's posture, and the head on top of the spine instead of in front of it reoriented the animals' perspective and extended their manual abilities. Not only were monkeys now able to manipulate stone and wood, they were now able to touch, groom, and embrace each other, which resulted in advanced social processes.

Until the advent of *Homo*, brain size remained barely larger than a chimp's. While brain size, measured as the cranial capacity of the skull, is a good measure of comparison *between* genera, it is not informative *within* a species. One method of calculating cranial capacity, a measure of braininess determined in cubic centimeters, is by putting small, standard-sized beads into the skull cavity and determining how many can fit. Chimpanzee cranial capacity averages about 360 cm³, the australopiths (an extinct cousin to modern humans) around 400–550 cm³, and *Homo erectus* up to 1,100 cm³. Modern humans measure about 1,350 cm³. The brain-to-body ratio—that is, how much brain tissue is available to run the organism—is a more important factor. Neanderthals also had a large brain, larger than the contemporary *Homo* species. Brain *size*, however, is not the whole story. Elephants, after all, have huge brains, but their brain-to-body ratio does not exceed that of the little primate tarsier, whose ratio is 1 to 26.

The major differences between our brains and those of chimpanzees are threefold: the development of the frontal lobes, the amount of association cortex, and brain lateralization, that is, the fact that the left and right lobes govern very different activities. Together, these major differences account for why certain skills, such as planning ahead, are so well developed in humans and not in nonhuman primates. This is especially true of the aptitude to take unrelated bits of information and put them together, creating a new idea. The ability to synthesize and construct is basic to the intellectual and aesthetic abilities that permit what we term arts and sciences.

Over the course of primate evolution, the frontal lobes expanded and expanded. While the shape of the brain is roughly like that of the chimp, the convolutions are quite different. A professor of mine, Harry Shapiro, used to expound on the fact that over evolutionary time, the brain could not continue its growth trajectory or it would grow straight out the back of the skull; it had to "roll up" to maintain balance on a bipedal strut, and, to further increase its surface area, it developed these convolutions. The more convolutions, the more surface area on these folds. It is these foldings, or sulcal patterns, that distinguish between humans and chimps. Here is where information is brought to the fore and where various bits of information can all be dealt with at the same time, integrated, and therefore acted upon. Part of the story is merely quantitative: we can bring *more* information to bear and dredge up more memories so that more information can be synthesized. The human brain is uniquely organized to deal rapidly with information that must be kept in sequence in order for actions to occur; to—at the same time—review what has been done and what is about to be done so that the future act can be corrected before it is set in motion. Timing and the *developed* concept of time are uniquely human. There is immediacy to the lives of monkeys and apes. Monkeys may dredge up a memory, which then becomes an active in-the-present stimulus to which the monkey responds. They appear to anticipate a climatic event, such as fruit ripening a distance away, or the Levanter (the eastern wind) coming, with its chills and misery, and respond *in advance*. In reality, however, they are receiving subtle cues—air temperature, turbulence, scent—which are stimuli in the present. Chimps do plan to some extent, fashioning tools for termite hunting at quite a distance removed from the termite nest; they also plan their attacks on neighboring groups, gathering themselves and marching "off to war." But apparently the extent of this ability is limited, at least in comparison to humans.

The left frontal lobe also houses the area responsible for speech. It is known as Broca's area and lies close to the areas for motor movements of the larynx, tongue, and the mouth itself. The other component of speech—understanding—is centered in Wernicke's area, in the temporal lobe of the brain near the primary auditory area. An arch-shaped part, the *arcuate sulcus*, joins the two areas. Rilling and colleagues (2008), with the assistance of new imaging techniques, have been able to pinpoint the essential differences between ape and human brains. Fossil evidence for these areas comes from the study of endocasts and dates to about 2 mya—not earlier. Endocasts are made by pouring latex into the bony skull. All the marks of the skull, including capillaries, bumps, nodules, and little holes are molded into the rubber, permitting comparison of one fossil to another. Monkeys and chimps process sound and meaning in analogous areas of the brain, hinting at what, from hindsight, became the future.

The chimp lineage went on to adapt to life in the trees, including eating, sleeping, and moving. Terrestriality was an important part of their adaptation as well and is particularly notable in the special structures of the wrist that permit knuckle walking. Relatively small chewing teeth with an elongated jaw in a face that projected forward beyond the brow ridges necessitated massive neck and jaw muscles. Points of attachment on the skull for these muscles stimulated growth into crests. In males, large upper canine teeth honed themselves against the lower premolars. The complex of traits useful for their adaptation included forelimbs adapted for knuckle walking to compensate for the huge bulk of shoulder, neck, and jaw muscles that were needed to move on large branches or on the ground.

It is important to note that growth rates differ significantly between humans and chimps, especially life-history stages like infancy, childhood, and adulthood. Infancy and childhood are less differentiated than adulthood, which sees the full expression of species-specific genes, so young chimps look more like us than do their parents. Those distinctly chimp traits—the huge neck musculature, the arms and hands, the pelvis and feet—do not appear until adolescence. The eruption of teeth is used as a marker of other life events. For example, 90 to 95 percent of brain growth is complete when the first permanent molar erupts. In a recent and exciting study, Zihlman and coworkers compared *wild* chimpanzee teeth to those of captive chimpanzees, and both to fossils, in particular *Homo erectus*, the species directly ancestral to us (Zihlman et al. 2004). Not only did the growth rates of wild chimps' teeth differ from those of captive chimpanzees, but the growth rates of the wild chimpanzees' teeth resembled those of *Homo erectus*. In both wild chimpanzees and *Homo erectus*, the first molars appear at close to four years of age, and the second molars at about eight years of age. In captive chimps, the timing is different as they mature more quickly, a sign of domestication. The significance of this is that our view of the growth patterns of hominins may need to be reevaluated (Zihlman et al. 2004). Part of the process of domestication is relaxation of selection on some genes, so that a variety of traits (molecular, genetic, anatomical, and behavioral) will thus appear different from the "wild" form.

While contemporary apes may be models for the beings near the time of our human ancestors, they are not to be taken as *the* ancestors. The apes' ancestor and ours would have been in the same group, of course, and that being would have been some sort of an ape, but it would not have been a modern chimp. There is a significant fraction of our genome that is more closely related to gorillas, and, as might be expected, there are portions of the chimp and gorilla genomes that are far more closely related to each other than either is to us. This is because gorillas and chimps separated from each other within a short space of time. Genes in the first ancestral species that gave rise to the gorilla-chimp lineage survived until the second divergence between human and chimpanzee lineages (Paabo 2003). This dating is confirmed by another source: the speciation of lice. The ancestral louse lived on gorillas, chimps, and humans. Somewhere around 13 mya, the gorilla louse remained true to and on gorillas but became extinct on humans and chimps. The lice that humans and chimps have, however, only speciated into different types around 6 mya—as their hosts were diverging. Humans "regained" the gorilla louse recently in time—around 3 mya—and the circumstances in which this occurred have not yet been determined (Reed et al. 2007).

A recent find in Spain of a creature quite clearly an ape has been dated to 13 mya, relatively close to the time of divergence of apes from the lesser apes (the gibbons, siamangs, etc.) (Moyà-Solà et al. 2004). Named *Pierolapithecus catalaunicus*, it is a good candidate as an ancestor to the group I am discussing, perhaps representing the last common ancestor, because it already intimates vertical orientation with no specialization for knuckle walking or suspension in trees. Its hands are like those of macaques or baboons, very much like ours. This is a spectacular find. There are currently very few "good" fossils from this time period. The features that mark it as a "primitive" form include traits seen in monkeys, like the slope of its face and short fingers. While it may have had an upright posture—and there are adaptations to suggest it—it also has some quadrupedal features in the way that monkeys have, putting their hands flat against the surface they are walking on. *P. catalaunicus* might have weighed 35 kg (77 lbs) and stood slightly smaller than a modern chimp. The apelike features include its ribcage, the discovery of which is a first in human paleontology for this age period. Similar to that of contemporary apes, the ribcage is wider and appears flat when viewed from front to back. The lower spine is not curved, but is straight—an adaptation for life in the trees.

The rectangular rather than arcuate jaw is a further diagnostic feature, and it holds large and impressive canines. Once again, a diagnostic feature—the shoulder blades—show this fossil to be closer to apes than monkeys since they lie along *P. catalaunicus*'s back, as they do in apes and humans, rather than being on the side of the ribcage (Moyà-Solà et al. 2004). *Pierolapithecus catalaunicus* was rather more upright than quadrupedal, a sign of vertical orientation in the trees. It certainly doesn't seem to have done much suspension from trees—its short fingers make that clear—climbing, yes, but hanging, probably not. Mind you, the proportions of a monkey's hands are far more like ours than are those of the chimps. The lack of apparent tree-climbing specializations suggests that contemporary apes developed their specializations more recently in time.

Some Thoughts on Charting the Course of Human Evolution

Contemporary ape forms are identifiable as such in the Miocene, the gibbons at 19 mya, the orangutans at 14 mya, the gorillas at 9 mya, the chimps at 8 mya, and the bonobos are discernable from the common chimps at about .86 mya (Paabo 2003; Won et al. 2005; Kunimatsu et al. 2007). But the period in which a species is first identifiable does not necessarily coincide with its first existence—the first descendant would look an awful lot like its parents, at least in some features. What's more, just as grandchildren live at the same time as their grandparents, each successive appearance does not mean the former incarnation has disappeared, much as we still inhabit the planet with gorillas (for the time being, at least).

Our view of human evolution used to be relatively simple, with a couple of genera as candidates for the different ancestors. Then more and more fossils were found, and the number of genera began to proliferate. The criterion to establish a new genus is that, ideally, a genus contains species members that share adaptations to a given environment. As renowned paleontologist Steven Jay Gould noted, the origin of a type sees a multiplicity of forms, as if nature were experimenting with what works (Gould 1989). The middle period of human evolution, clustering above 4 mya and continuing to about 3.5 mya, sees this kind of diversity in genera (Leakey et al. 2001). The variety of forms raises the question as to whether or not some of these might be ancestral to chimpanzees rather than to our own line. Most of the fossils in this time period show a mosaic of traits. Some features they exhibit seem to be shared with an earlier form, while others are "derived"—totally new in this group as compared to an ancestor—but possibly shared with another descendant. These composite forms were at first perplexing as it was assumed that a clear differentiation between fossil types would be the norm. Now we know differently, and we expect that as times and conditions change, a multitude of new forms will emerge at approximately the same time, as if testing various features to enhance adaptation. This "experimentation" is termed an *adaptive radiation* and seems to have occurred at several points in hominid history. The recent findings with regard to the role of tandem repeat sequences suggest one mechanism by which this works.

Some years ago, Gould noted in his book on the 500-million-year-old Burgess Shales of Canada that each adaptive radiation is marked by an abundance in types of organisms exploiting this new way of life. Rather than a ladder of life forms, evolution is evident in this rich, branching diversity. Because the anatomy of the different kinds of organisms that live on this planet has been very well known for centuries, placement of fossil materials into a *clade*—a lineage of ancestor-descendant forms—is possible. Graphic depiction of evolutionary "trees" show branches and twigs shooting off in different directions with dashed lines indicating where the fossil history—the actual feathers, teeth, or bones—is missing. The general outline can be known, but the details are still being filled in. As I write this, a 375-million-year-old fossil has just been found that bridges the gap between water animals with gills and fins and land-based animals with lungs and limbs. The *process* of transformation was known, but this fossil confirms the time and means by which this major evolutionary event took place (Shubin et al. 2006).

It is more accurate, therefore, to think of our genome as a mosaic, with different histories for different segments. The *expression* of genes is what is important; the genetic *material* may be similar to ours, but the *reading* of these genes differs importantly, so that the end products are distinct.

The Human Ancestor and the Role of Bipedalism

The Ancestor, then, a generalized ape in the process of becoming human, gradually colonized a terrestrial habitat, considered now to be woodland with clusters of trees, shrubs, and grasses—not savanna—by means of bipedalism, although retaining tree-climbing abilities. If the 13-million-year-old ape recently found in Spain is typical of the last common ancestor, the Ancestor would have had hands like monkeys, which are proportioned like ours. The specializations in the hands of today's apes would not yet have appeared. The Ancestors closer to 6 mya certainly had all the other attributes of apes, including their cognitive abilities, although they were probably a bit smaller than a chimpanzee.

The Ancestor would have shown some signs of hominin status, especially with regard to habitual upright posture and the ability to walk bipedally. The Ancestor's chewing "apparatus" would have combined proportionally larger chewing teeth with canines in the male that were rather small in size but that wore only at the tips, as do ours. Female teeth were smaller, in proportion to their size. Their brain would probably have been about the size of a chimpanzee, around 450 cc. Recall that differences in brain size *within* a species is not indicative of ability; *between* species however, size has its implications. One of the smallest totally human brains measured a mere 1,000 cc, where the average is closer to 1,400 cc. (It belonged to the French author Anatole France whose book *Penguin Island* was both a sensation and delight.)

Modern humans can also be quite adept at tree climbing but tend to assist the limbs with ropes, cleats, or other devices. I have seen photos (at the American Museum of Natural History) of Tibetans that showed lifetime alterations in their feet due to clambering over rocks for their livelihood. Over time, this made their feet better clinging instruments, but, of course, their children had feet like the rest of us. In the same way, Pacific Islanders can climb palm trees without benefit of ropes or the like because their feet become splayed and able to grasp. This is not a genetic trait, however; rather, it affirms the malleability of bones and joints. The continuing ability to adjust to tree climbing reflects the flexibility of our species.

The ability to walk on two legs, however, is the truly difficult feat. The bipedal walk has two phases: the stance and the swing. Just standing on two feet in humans requires no muscular energy. The shape of the pelvis, the angle of the femur (thigh bone) as it contacts the knee, the shape and position of the head of the femur as it articulates into the hip bone, and the curves in the spine all permit passive balance (Alexander 1995). Walking requires dynamic balance that depends on these anatomical features and a complicated interplay of muscles that subtly shift the weight from one leg and foot to the other while the body is moving forward. These oscillations between kinetic and potential energy permit conservation of 70 percent of the energy involved (Sellers et al. 2005). The basic anatomy of the femur is a shaft ending in articular surfaces at the knee and hip. At the upper end, a piece of bone, the femoral neck, unites the shaft with the ball or head of the femur. This ball fits into the socket at the hip and is the one that permits the various movements of the leg. Length and thickness of the femoral neck are exceptionally revealing. In contemporary African apes, the thickness of the upper and lower parts of the femur are about the same, but in our species the ratio of thickness top to bottom is one to four, or higher. The femoral neck is not very long in apes, but it is in humans.

In a quadruped, the vertebra where the spine bends like a bow is called the anticlinal vertebra. The center of gravity lies beneath this. In bipeds, the center of gravity descends down to the bottom of the vertebral column and along the outside of the knee, whereas in the knuckle walker, the center of gravity is somewhere around the midsection. The spine of quadrupeds and apes is straight—from head to bottom. In humans, there are two forward and two backward curves, the first in the neck, with the curve just below the neck stabilizing the area, the lordosis in the lumbar region, and the gentle curve outward at the coccyx or tailbone. The number of lumbar (small of the back) vertebrae in modern humans is five, in chimps four, and in early hominins, for reasons not yet understood, six (Filler 2007)! Interestingly, the muscles of the back permitting bipedalism have been found to be similar in humans, gibbons, and chimps, but the "architecture" and mechanical advantages of these muscles have improved over the evolution of this gait (Shapiro and Jungers 1988). The bowl-like pelvis is intrinsically stable, with *two* forward curvatures in the spine, particularly the lumbar, which serves as a counter balance, and thigh bones that angle down to the knee transferring weight in a straight line through to the foot so that weight falls between the big toe and the second toe. The hip bone, providing the socket for the head of the femur is a three-dimensional structure. The inner part of that bone contains spongy tissue, likened by some to a sandwich, in which the inner portion of bone, known as cancellous tissue, acts as a spacer to the compact (denser) bone on the outside. This structure disperses the stresses of the bipedal stride. Like the lordosis curve of the small of the back, the cancellous matrix in the hip bone is not well defined at birth. As toddlers begin to walk, little spikes of bone—the trabeculae—thicken within the matrix, strengthening it. In the 1960s, a professor of mine, Professor Riesenfeld, who studied experimental anatomy, transformed a quadrupedal rat into a biped, complete with that lordosis (Riesenfeld 1966). The recent studies of Japanese macaques affirm that bipedal posture is not only attainable by mammals other than humans, but in so doing the lordosis forms distinctly. However, the gait in these experimental animals is not the true gait of a habitual biped (Hirasaki et al. 2004). The conformation of the pelvis, especially the iliac bones, means that the origin and insertion of muscles is never going to permit the stride-through that is characteristic of humans.

Most importantly, perhaps, is the shape of the basin formed by the three pelvic bones: the ischium, the ilium, and the pubis. The ilium is a long, narrow bone in apes, but in committed bipedal humans, each ilium has broadened and rounded to become the side of the pelvic basin. Muscles insert into bone. In apes, the gluteal muscles extend along the back and insert into the long, narrow ilium, assisting in propelling the ape up a tree. The broader expanse of the ilium in bipeds offers more gripping area for the gluteal muscles that stabilize the leg, and the muscles run along the back of the thighs and down the sides. The "glutes," as exercisers call them, are the muscles that shape our bottom end. While humans may have an advantage in that they can walk upright, human bones are less mobile than those of chimpanzees. This is the tradeoff for bipedalism, where the lower limbs support all the body's weight (Kunimatsu et al. 2007).

Bipedalism is rare among animals, although kangaroos, an example of pouched mammals, are bipedal with the aid of a powerful tail, and bears, an example of placental mammals, are often found standing on two legs. Birds, of course, are two-legged. But *walking* on two legs rather than shuffling, hopping, jumping, or some other gait is a human trait, and it was therefore always assumed that this locomotor innovation would coincide with a creature far more like us and closer in time to us. Walking on two legs means that the hands are free for carrying objects. The eyes are forward and above the body, the gaze moves freely across the horizon, and the body easily twists to look behind one's self. Yes, chimpanzees do carry objects as they move on three limbs, but the size, kind, and even the weight of these objects will be limited by a three-limbed gait. The tendency toward bipedalism goes back to nearly 20 mya (Filler 2007; Maclatchy 2004). Anatomical evidence suggests that the *ability* to move on two legs preceded the *requirement* to do so, when anatomy dictated this obligate form of locomotion.

Oddly, recent evidence suggests that the special adaptation that knuckle-walking apes have in their wrists was present in some of our more recent ancestors. I say "oddly" because a general rule of evolutionary biology is that a specialized adaptation does not permit going back from that adaptation. The heavy neck, jaw muscles, and huge shoulders would not easily balance on a spine like ours, having only two forward curves, and given the narrow pelvis and legs flared to balance. The issue of the wrists is a vexed one; we do not have sufficient postcranial fragments of early hominins from below the head to determine what the wrists were like in the earliest fossils. Then, too, the specialized wrists come from the middle time period, around the 4-million-year mark. The marvelous aspect of science is that knowledge is constantly changing and disagreements engender new research. In the case of the wrists, there are distinctly two groups of researchers: those that see the fossils with specialized wrists as representing a stage in the evolution toward humans and those that see this specialization as clearly marking the path toward contemporary apes. After all, the ancestor of us both was not yet human or ape!

The wrist of *P. catalaunicus* (13 mya) seems to be midway between humans and apes and indicates specialized climbing abilities that link it with modern great apes. It is joined, however, to only one of the two forearm bones, which permits rotation, great flexibility, and a configuration similar to ours. The wrists, with their specialized bones, can take the combined weight of shoulders and head in forward locomotion. But such a specialization is hard to retreat from; the shortened forelimbs and lengthened hind limbs of bipedal posture are a distance from the long arms and short legs of committed knuckle walkers. I must confess I side with Professor M. Dainton in this. His recent analysis of the wrist bones in fossil ancestors led him to conclude that those fossils were not generalized enough to be a common ancestor for chimps *and* humans (Dainton 2001). Ape specialists also concur. In his book on bipedalism, Craig Stanford (2003) notes that there is considerable support for the idea that knuckle walking in the African apes came *after* the divergence from the stock that gave rise to humans. Hence, the specializations for knuckle walking, including intimate anatomical details in the wrist as well as the ridges on the skull for the huge muscles of the neck to insert, and all other adaptations, including leg-to-arm ratios, would be relatively recent and not related to the development of human locomotion.

Why Bipedalism?

Why bipedalism arose remains speculative. There are a variety of popular reasons suggested, some of which seem more fiction than fact (see Rose 1991 for a complete review). Field workers studying nonhuman primates have long been aware that when the animals have something they want to carry, they stand and shuffle-run bipedally. I have seen baboons in Kenya raiding a local field for watermelons and either tucking one melon under each arm or running off with a big melon across both arms and under the chin, tails waving crazily as they bounded away. I have seen an adolescent male macaque, having kidnapped a baby from its adult *male* caregiver, run off with it bipedally, indifferent to the squealing youngster held topsy-turvy, feet up and head down, against its chest.

Chimps and bonobos are far more adept at bipedalism and resort to this posture more frequently and casually, often motivated by wanting to carry something. For these apes, it may be a twig with which they capture termites, a stick to be fashioned into a tool, or a vine to be draped over the body as apparel or decoration. Perhaps this was enough of a reason for the Ancestor as well. Another suggestion is that many mammals that can sit upright, like squirrels, for example, will stand from the sit position to look about. Prairie dogs do this, and so, of course, do nonhuman primates. This *vigilance* behavior encourages upright posture; whether or not it also impelled upright locomotion is another matter. I personally doubt it because the frequency and duration of vigilance bouts does not seem to be enough of a selective pressure. A similar notion is that feeding in bushes and off the lower branches of trees encouraged bipedality. The more fluid the upright posture, the better the accumulation of foods, and hence the inclination to a locomotory system that is dependent on upright posture. Experimental evidence confirms that carrying objects and standing upright to forage were indeed activities that fostered bipedality (Videan and McGrew 2002). Hirasaki and colleagues (2004) worked with macaques trained to walk bipedally and found that training itself considerably improved bipedal walking, suggesting that the earliest bipeds who were not yet fluent in that gait became more adapted as they continued to walk bipedally.

A new idea comes from observations of social behavior, particularly communication, in monkeys and chimpanzees. During dominance confrontations, and occasionally in sexual displays (at least in bonobos), an upright posture permits the incorporation of noisy or threatening objects like branches in the display. Baboons and macaques will raise themselves up onto their hind legs (as do bears) in agonistic encounters, as well. This type of behavior, if present in early primates, may have been a harbinger of things to come (Stanford 2003).

Certainly, bipedalism does not seem more efficient (Steudel-Numbers 2003), inclining scholars to consider other factors—thermoregulation and energy efficiency, for example. The vertical orientation of bipedalism would help cool the body as less surface area is exposed to ultraviolet rays (Wheeler 1991). There are variations in a gene named *MC1R* that provide pigmentation as protection against UV light. These variations would have permitted the hairless trait to expand through the population (Rogers et al. 2004). In addition, height from the ground might expose the body to cooling breezes and may also explain hairlessness that promotes evaporation of moisture from the skin; and this, in turn, selects for pigmentation to protect the skin from UV and skin secretion of melatonin (Fischer et al. 2006). Ectoparasites, clinging to hair as they do, are considered by some to have encouraged the evolution of hairlessness (Rantala 1999). As fire became increasingly indispensable, selection *for* fur would have been relaxed and hairlessness would have been a more frequent variant. The Yaghan, traditional peoples of Tierra del Fuego, Argentina, serve as an example of people with scant clothing and notably bare feet in a landscape that becomes extremely cold even before the sun goes down. Fire was their only source of heat and was even brought into their canoes. They had fires burning constantly, to the point that the early Spanish explorers named the area Tierra del Fuego, or Land of Fire. In Australia, shelters were often constructed by the Aboriginals for just one night as the group traveled from area to area. The group lay around the fire such that everyone could get some heat reflecting onto at least their backs.

Experimentation on walking has demonstrated that it requires less energy to walk upright than to run. A 70-kg (154-lb) man, for example, uses 140 more joules per meter (33 calories) when walking than when standing still. Running requires 260 more joules (62 calories) than does standing still. Quadrupedalism is still more costly. A mammal weighing as much as the person in our example would use 200 more joules of energy per meter (48 calories), whether walking or running (Videan and McGrew 2002). Indeed, the ability to *run* came considerably later than the ability to walk—which in any case was not perfected for millions of years (Sellers et al. 2003). However, the impetus for the initial or *germinating* species (*species germinalis*—that is, one where all diagnostic features of that species may not be present, and some traits may more clearly resemble an ancestral form) would probably not have been energy efficient; the transitional member is invariably a mosaic of traits between "what was" and "what is becoming" and would probably not have been any better at walking bipedally than a very gifted chimp. As this form of locomotion became standard, when forests gave way to woodland and then savanna-grassland, selection, especially of the lumbar curve, would have been geared toward it. The Ancestor—without shelters and not yet manufacturing fire for protection—would have gone from tree cluster to food source and then back again to sleeping trees over the course of the day. This behavior explains the anatomical features that suggest that tree climbing was still a habitual part of the Ancestor's repertoire and also helps to explain the motivation for walking. Evidence from fossils from up to 4 mya shows varying kinds of bipedality, with what was probably varying degrees of efficiency. By the time of the genus *Homo*, toward 2 mya, the posture and gait were a little different from ours. Certainly the erect posture and bipedal locomotion selected for further refinements in posture and locomotion.

Bipedalism and Evolution of the Human Brain

The "privileged evolution" (as geneticist Bruce Lahn calls it; see Evans et al. 2004) of the human brain was rapid and is thought to have depended on bipedal posture. As early hominins moved erect, the brain, now directly under the sun and no longer as protected by living in the forest, required special apparatus for cooling. Dean Falk has discussed this at length, noting that the "modern" hominin pattern seems to begin with australopiths and is realized in *Homo* (Falk 1992). While other mammals have a network of arteries and veins that help cool the brain, humans do not. We have a *vertebral plexus* of veins around the spinal column that collects considerably more blood when we are standing up as opposed to lying down. In addition, the sinuses in the brain and other circulatory structures are unique in humans, and, even more than our anatomy, our *genetic* activity is distinctly different from that of chimpanzees, as Svante Paabo and his team have demonstrated (Evans et al. 2004). Interestingly, our liver and other tissues have similar genetic activity to these chimp cousins, while differing—as we might expect—from the macaque.

Our brain is a hungry organ. It requires vast amounts of energy to maintain itself (Leonard and Robertson 1992, 1994; Paabo 2003). Indeed, some believe that bipedalism evolved in part because it is less expensive, in a caloric sense, than quadrupedalism. Contemporary human foragers typically walked up to 4 km (2.5 mi) a day in search of food, whereas apes move in search of food over perhaps half a kilometer or so (Leonard 2002). The energy required to maintain our modern brains amounts to as much as 25 percent of our energy needs, whereas most mammals require only up to 5 percent, and primates, so much closer to us, need 10 percent at the most. Much of our energy needs are met by high-quality foods, especially in the form of animal protein. Chimps take relatively little animal meat— about 5–7 percent of their energy needs—although they do hunt, but do not scavenge (Ragir et al. 2000). Eighty-five grams (3.5 oz) of animal protein provides around 200 kilocalories. Compared to plant matter, this is high-quality food indeed, as a comparable amount of fruit would yield half the kilocalories, and leaves would only provide up to 20 kilocalories. Leonard and coworkers consider that the human brain was able to grow in size over the eons due to our increasing dependence on animal protein.

The australopiths had apelike brains at a capacity of around 400 cc (at 4 mya), while the brain of *Homo* quickly expanded from 600 cc in *Homo habilis* (2.5 mya) to 900 cc in *H. erectus* (2 mya), just under the lowest human value (950 cc). This brain growth, however, was not yet seen in the time of the Ancestor, whose brain was still chimp-sized (Leonard and Robertson 1992, 1994; Leonard 2002). It may be easy to say, but it is likely that more than one of the factors suggested provided the impetus to habitual bipedal locomotion. Upright posture begins with sitting erect. Perhaps the most significant evolutionary change in our line preceding bipedalism was the freedom of the forelimb, especially the hand, and the concomitant changes in spine and cranium that went with it.

Fossil Evidence of the Chimp-Human Divergence

Fossils have been found as early as 6 mya in Chad, and just a little later in Kenya (i.e., closer to now), but whether the earliest are chimps and therefore apes, or whether they are forms on the trajectory toward modern humans, is not yet clear. Several features have been identified as diagnostic of the human condition, and these are applied to these ancient fossils to help sort out the issue. Habitual upright bipedalism as the basic method of locomotion is a defining characteristic; others include a face that comes under the brows, a jaw which is more semicircular or arcuate than rectangular, and certain features of the teeth, including, for example, thickness on the enamel of molars, and more importantly, small canines that are more like incisors in form and that wear at the tips as do ours.

Yet at these early time horizons, when chimp and human were in the process of diverging from each other, there is often ambiguity in the features. The earliest hominin forms for which there are fossils are given in table 5, including the age, where it was found, what was found, and whether or not the remains have been interpreted as bipedal. The names of the fossils at the time of divergence follow the Linnaean rules of nomenclature: genus first and always capitalized, species second and always lowercased. *Homo sapiens*—our group—is the best-known example. The latest fossil names, however, represent the newer trend to utilize local terms or place names with respect to the locale and even the language of the place in which they were found.

As one might imagine, the discoverers of the respective fossils vie for primacy in establishing their fossil as the type from which our species is ultimately derived. Each discoverer finds interpretative explanations as to why the other fossils are ancestral to some other form, or not at all involved. We know that *Sahelanthropus tchadensis* (Chad) lived between 7 and 6 mya (Brunet et al. 2002). However, the sediments in which the various pieces were found cannot be dated by the absolute methods permitted by establishing the rate of change between radioactive and stable forms of potassium and argon isotopes. Furthermore, information on changes in the earth's magnetism, which provides a reliable dating of the earth, is not available for this region. In this case, it was therefore necessary to employ the comparative method, which evaluates the flora and fauna of sediment and compares it to another area where the dating of the same

Table 5. Fossils and their details				
Name	Location	Age	Found	Biped?
Sahelanthropus tchadensis	Chad	7–6 mya*	Most of cranium	Biped?
Orrorin tugenensis	Kenya	6.2–5.6 mya	Femurs, teeth part mandible	Biped!
Ardipithecus kadabba	Ethiopia	5.8–5.2 mya	Mandible teeth, partial hand bones, toe bone	Biped??

Note: * Unit is in millions of years ago.

sorts of flora and fauna is possible. The material from two sites in Kenya was compared to the one where *S. tchadensis* was found. Comparing new fossils to fossils of a known date follows the logic that similar forms of life in two different areas would have lived at the same time. Through calibrations such as these, the fossil from the Sahel (Chad) has been reliably estimated at between 7 and 6 million years old.

Some six specimens were unearthed from the Sahel, including a nearly complete skull and parts of lower jaws and teeth. Some of the traits are distinctly reminiscent of the creatures from whom *S. tchadensis* evolved. The small capacity in the skull to encase the brain, measured as the cranial capacity, is sufficient to house a brain at the lower end of the range for a chimpanzee. The assumption, therefore, is that cognitive processes would have been similar to these cousins—a flawed premise, perhaps. The front teeth are also similar—the central incisors are large, and although the canines are still fanglike rather than flat and short, they are not as large as those seen in apes. Most important are two features of the skull that suggest that *S. tchadensis* may have been bipedal. The position of the bony crest above the neck at the back of the skull (the nuchal crest) argues for muscular attachment permitting erect posture. It contrasts with the same area in apes, where the nuchal crest is powerfully developed to hold up the large, projecting cranium. The shape of the hole at the base of the skull, through which the spinal column reaches the brain, or *foramen magnum*, is not rounded as it is in chimps. Rather it is longer than it is wide. This is the diagnostic criterion indicating a head balanced on its spine and permitting forward motion without ancillary support. This evidence from the skull alone is not conclusive, however; limb bones will provide a definitive answer, if and when they are found.

S. *tchadensis* apparently lived in a swampy area near open grasslands and forest running alongside a river or creek, called a gallery forest. That environment is precisely the kind of mixed habitat that would most stimulate a flexible primate to use all its ingenuity to seek out its livelihood. Different zones produce different sources of food and, equally as important, different sorts of dangers. All of this information has to be evaluated and retained over generations to secure the fate of the society. S. *tchadensis* presents a *mosaic* of traits—some more like its ancestor and other new traits that are more frequent and developed in its progeny. The face juts forward less than it does in *Pan*, the chimpanzee. Yet the tall face has a massive brow ridge, and the braincase is long rather than rounded—a feature reminiscent of apes. It is reasonable that creatures evolving from one state to another will show such a mosaic of traits.

Recently, the term *species germinalis* has been applied to transitional species such as S. *tchadensis*, because some traits are more like the ancestral form and not all diagnostic features of that species may be present. As we discussed earlier in this chapter, decisions about how to categorize a fossil such as S. *tchadensis* reflect the current range of views regarding human evolution at its earliest beginnings. One is the linear model, wherein all the anatomy distinctive to hominins evolved only once, and the descent follows the ancestor-descendant series rung by rung on the metaphorical ladder. This view sees no branching of hominins—a process called *cladogenesis*—until after 3 mya. The opposing view accepts that evolution is not neat and that a branching image more graphically depicts what happened. This view sees humans evolving in a succession of diverse forms in response to changing conditions. Such adaptive radiation underscores the flexibility in primates generally. Here, mosaics of traits are frequent, as the development of different kinds of beings at similar times is rather like a multitude of experiments taking place at the same time. Adaptations particular to hominins—bipedalism, the large brain, the small dentition—seem to have appeared more than once in different lines. The difficulty arises, then, in linking each fossil to a subsequent form—to locate an ancestor-descendant grouping—and so simple a picture may not be accurate anyway. It used to be thought that the simple succession of fossils indicated a descent line. Now we know that this may not be the case. Certainly, in general terms, one form precedes another, but clusters of traits may be retained for millennia such that more "primitive" forms coexist with more "derived" types.

The implications of regular bipedalism, especially at this time horizon, are dramatic and also extend to control of body temperature, as more surface area is exposed. By 1.2 mya, humans were almost completely hairless (Leonard 2002). The process of becoming so coincides with shifting climate and therefore landscape, and, according to the scholars who have postulated this, is a result of several "sweeps" (where variation is decreased) occurring in a particular gene. The process may well have begun around the time of the bipedal Ancestor, and the relationship to fire would have assisted the genetic process because it would have substituted for fur in protection from stings as well as for temperature control. In any case, loss of body hair did not occur in the armpits, pubic zone, legs, arms, face, and, above all, top of the head. Body hair elsewhere is not really even lost—it simply does not develop. The number of follicles humans have is actually very close to what is seen in the great apes.

Bipedality in *S. tchadensis* is assumed on the basis of comparison to the features in the limbs of another Ancestor, *Orrorin tugenensis* (Senut et al. 2001). Dated at 6 mya, *O. tugenensis* was found in Kenya. Many traits in this Ancestor suggest a direct descent to the genus *Homo*, of which we are a member species. Once again, the model of how human evolution took place is more at issue than is the placing of the fossil. Is *O. tugenensis* a member of the adaptive radiation or an ancestor in our own lineage? The small teeth of *O. tugenensis* are suggestive of a more direct relationship to *Homo*, as is the thick molar enamel—an important diagnostic trait. The African great apes do not have thick molar enamel; it is considered a hominin trait (Vogel et al. 2008). Five individuals were in the sample of some twenty fragments of bone. The late Miocene date of approximately 6 mya is firm, based on biostratigraphy, geology, radiometry, and paleomagnetism (Haile-Selassie 2001). There were only three pieces of thighbone, one of which had an intact head. Analysis of these suggests that bipedality may already have been an obligate behavior, a restriction of its anatomy, compelling *O. tugenensis* to walk on two legs (Galik et al. 2004). The difference this makes is enormous, especially at this time horizon. There are several diagnostic features on the thighbone that reveal whether or not a fossil ancestor was capable of bipedal walking. Researchers examined the fragment of *O. tugenensis* with shaft, neck, and intact head. Although the femur was similar in size to that of a contemporary chimpanzee, the three-dimensional image of the CAT (computer-aided tomography) scan showed that the femoral neck between the ball and the shaft was thinner on top. This comparison to fossil apes confirmed that *O. tugenensis* was unlike them and clearly had to be grouped with hominins because the ratio of upper to lower thickness in *O. tugenensis*'s femur was 1 to 3, quite close to the human value. The long femoral neck in *O. tugenensis* indicates that gluteal muscles functioned then as they do now in humans during the pelvic-support phase of bipedal walking (Galik et al. 2004).

The last fossil member in this early time is *Ardipithecus kadabba* (Haile-Selassie 2001). This is one of two species of *Ardipithecus ramidus* found in Ethiopia (some scholars consider these two forms as subspecies of *ramidus*; that is, *A. ramidus kadabba* and *A. r. ramidus*). Its name in the Afar language means "base family ancestor." So far there are some eleven specimens from five localities dating from 5.8 to 5.2 mya. The finds are fragmented but include parts of the most important bits of anatomy. Together they begin to tell a story about this being. What has been found to date is the lower jawbone complete with some teeth, part of a collarbone, some hand bones, and most important of all, a toe bone. These permit some reconstruction of diet, posture, hand use, and locomotion. Based on the evidence of that toe, some believe that this nearly 6-million-year-old fossil may have been bipedal. Other authorities, however, disagree. Interpretation of the bones and their features is contentious. Those finding other near-contemporary fossils suggest that these remains describe a chimpanzee, so the position of *A. kadabba* in the panoply of ancestors is uncertain. However, the balance sheet favors a relationship to hominins with several traits suggestive of this heritage. The wear on the top of the lower canine is like that seen in *Pan*, the chimpanzee, but there is a difference in the details. Whereas apes usually wear this part of the tooth in a diagonal direction, in *A. kadabba* it was worn horizontally. The minutiae are significant as they indicate that the fully functional honing system that is formed by upper and lower canines wearing against each other, and which is found in primitive apes, was not present in *A. kadabba*. And, to the contrary, the incisorlike shape of the lower canine is a distinctly hominin trait.

One of these five *Ardipithecus kadabba* individuals suffered from periodontal disease, known from an abscess and consequent swelling (Senut et al. 2001). It has always been assumed that our ancestors, and indeed our cousins the living great apes, could live an entire life free from these sorts of impediments. Not so! Broken bones, cracked teeth, gum inflammations, and the like were as much part of their world as ours. Adolph Schultz was the first anatomist to recognize that tree-adapted primates were as klutzy as the next guy, and they paid for it, too (Schultz 1969). He found that apes as well as monkeys often died from their mistakes in arboreal locomotion. Undoubtedly the Ancestor was no less liable to physical injury. The use of plants as medicine probably has its origin here as well, as the Ancestor turned to Mother Nature for help in battling disease. After all, nonhuman primates use herbs for everything from cleansing wounds, to removing parasites from the fur, to easing tummy troubles (Huffman 1997). Even more impressive than the usage of this pharmacopeia, they seem to know the toxic limits of the materials they use and ingest (Engel 2002). I saw female monkeys in Kowloon using hibiscus flowers, which local women eat when they have mastitis. The problem was that I could not ask the monkey if she had a problem. Less circumstantial was the observation of monkeys eating licorice plants to expel afterbirth, which was observed by the keeper in charge of apes in Gibraltar (Burton 1972b).

Some of the foot bones found at 5.2 mya are considered derived—that is, having new features relative to all known apes—and suggest what the paleontologist considers an early form of bipedality. The finder of *Ardipithecus* (both *kadabba* and *ramidus*), Dr. Yohannes Haile-Selassie, draws our attention to the sequence of fossils that have been found dating from the 5 million years following *Ardipithecus* in the same region. He notes that the sequence favors interpreting *Ardipithecus* as a hominin and related to further human evolution (Haile-Selassie 2001). Remember—the fact of human evolution is not the question, nor are the stages of hominization. What is controversial and not certain is identifying which fossils form the ancestor-descendant line leading to modern humans. Certainty, or consensus, at least, appears around 2.5 mya with the earliest members of our genus, *Homo*. There are several species within the genus *Australopithecus* dating back to between 3.5 and 4 mya that are candidates. Which ones actually belong in this genus, however, and what their relationship is to *Homo*, is now under review. The finds in the Awash Valley of Ethiopia that cover the 5 million years after *Ardipithecus* include several types of *Australopithecus*. The genus *Homo* appears close to the 2-million-year mark.

Sahelanthropus tchadensis, *Orrorin tugenensis*, and *Ardipithecus kadabba* lived in environments that some few decades ago would have been considered unlikely homes for the Ancestor. It used to be an accepted notion that our ancestors had to have lived in dry zones, which, based on the model of baboons, would have encouraged upright posture for several of the reasons outlined earlier. However, environmental evidence for each of these fossils suggests that over the past 6 million years environmental conditions were quite variable in Africa, where peaks of moisture and vegetation changed to troughs of dryness and arid conditions. The variety of environments our ancestors would have lived in encouraged selection for behavioral adaptability as well as physiological variability. They could live in both densely wooded and open habitats—knowing how to do so was a result of accumulated knowledge over the eons, and, in turn, the area encouraged further innovation and adjustment. Indeed, for *A. kadabba* home was a wet, wooded area, while for *S. tchadensis* it was swampy, surrounded by woods and near to grasslands, and *O. tugenensis* lived in a wooded habitat. The hallmark of nonhuman primates is flexibility, and adapting to different environmental requirements fosters that trait. The implications for cognitive abilities, analysis, and memory are suggested by what we know of contemporary nonhuman primates.

Hominization and the Role of Self-Domestication

By the time of the third, most recent, period, somewhere around 2.7 mya, there was a rapid concatenation of events toward hominization. Stone tools are found at 2.6 mya; and by 2.5 mya, the stones indicate that the hominins were using foresight for later use in the "curating" of their tool cores and flakes (de Heinzelin et al. 1999). By this time, large mammals were being killed, dismembered, and stripped of their flesh to obtain what was a "new" food—the marrow in the long bones. It would have been a lot easier in the long run, and metabolically speaking, to dispatch a relatively large mammal and break open its bones for the marrow, which provided fats and protein, than to hunt for small vertebrates and insects. And, although evidence for fire is not well established until much later, 2.5 mya is suggestive of an earlier use, if not manufacture, of fire, the remains of which simply have not been found. The diversity of fossil forms at this time horizon is also suggestive: it means rapid change was taking place, of which at least some was under the impetus of the cultural processes accumulated to that date.

The diversity itself suggests that domesticating processes had *already* taken place. These occurred because of flaws in DNA copying of gene sequences. In turn these "errors" resulted in new productions, thereby diversifying members of a similar genome. *Homo* begins with *H. habilis* around 2.5 mya, followed by *Homo rudolfensis* and *H. ergaster*, at 1.9 mya, and *H. erectus* beginning around 2 mya. If *Homo* is already domesticated, as evidenced by the diversity in members of the genus just around the 2-million-year mark, then one important agent of domestication—fire—and its ancillary effects must precede this period by a considerable amount of time. Indeed, the earliest evidence of fire has now been established at 1.6 mya at Koobi Fora, Kenya (Bellomo 1994a, 1994b).

Earlier in this chapter, we discussed genetic drift as an evolutionary force in hominin evolution and noted the work of Ackermann and Cheverud (2004) in highlighting the random nature of patterns of variation in early *Homo*. It occurs to me that perhaps tandem repeat sequence evolution, also discussed earlier, might be a mechanism underlying this diversity. If that were the case, it would further support the idea that at least some species of *Homo* was well on its way to domestication. When diversity is caused by natural selection, the correspondence is clear between a feature and its function. For example, the chewing apparatus of the large and heavy robust australopiths may be the force selecting for large cheekbones that give large muscles the necessary surface to work over. What relaxes selection has long been thought to be *cultural processes* mitigating the need for physiological adaptation. Tools, for example, release the organism from the need for fangs or claws. Stone tools are evidence for the use of objects outside the body. The ancient Greeks, Darwin, and Kenneth Oakley, the physical anthropologist (see Oakley

1955), to jump centuries in a single sentence, considered that *man makes himself* (to use Childe's title of the book he published in 1965)—that the process of hominization is equivalent to self-domestication.

What sort of cultural structure must have been necessary for hominins to begin the process? Are the final stages of hominization equivalent to domestication? We could identify such a structure as a "home"—the domus, its central fire (and eventually a true hearth or even fireplace), and tools substituting for inadequate physiology. These are technological accoutrements accompanying, I think, a worldview. The epic of Gilgamesh encapsulates the history of the development of this worldview. In a nutshell, it is the story of a hero, Gilgamesh, who is warned of the prophecy that he must fight with the wild man—Enkidu. Gilgamesh already represents the city, while Enkidu is nature unspoiled. They wrestle and tumble, and ultimately Gilgamesh leads Enkidu to the city, Ur. He is treated to "a woman's task" and subsequently can no longer speak to the animals—he has been "civilized." This wonderful tale was written thousands of years before the Old Testament. The worldview represented here is of humanity "conquering" and displacing nature. This being sees itself as no longer subject to the environment; rather, humanity now sees itself as controlling all that is.

How does this begin?

In reconstructing the Ancestor of this earliest time horizon, much of what can be imagined is dependent on what is actually known of contemporary nonhuman primates, particularly Old World monkeys and apes. Since contemporary apes have their own, as yet not fully known, fossil history dating back at least 5 mya, what they do nowadays does not mean that that is necessarily what they did as they were becoming chimps. Asian and African monkeys, that is, Old World monkeys, appear in the fossil record before apes, although they do not proliferate into their various types until after apes have quite nearly gone extinct in the Late Miocene (Raaum, et al. 2005). With this caveat, however, let us turn, in the next chapter, to an examination of nonhuman primate behavior as we observe it today, especially with regard to their lives in social groups and to their cognitive abilities, and ask whether it can be used as a model for the emerging Ancestor.

Baby, Light My Fire

Apes and Old World Monkeys as Models for the Ancestor

We have already explored some of the ways (particularly in chapter 2) in which the behavior of contemporary non-human primates might inform our hypotheses concerning why and how the early Ancestor chose to *approach* fire, despite its obvious dangers and capacity to induce great fear. In this chapter, the idea is to further refine a mental image of what this Ancestor may have been like from what is known of the lifeways and higher-order cognitive skills—especially communication—of nonhuman primates in order to figure out how the Ancestor would have further developed the routine use, and ultimately domestication, of fire. In doing this, we will also touch on the still controversial issue of whether the lifeways and behaviors of contemporary apes and monkeys can be said to constitute "culture."

What would such a characterization imply for our developing scenario of the first use of fire by our Late Miocene Ancestor and its continuation on down through multiple generations, reaching ultimately to *Homo sapiens*? We have already established that some behaviors are universal to nonhuman primates, or at least, and more important to this book, universal to apes and Old World monkeys. Because they are found widespread in these two segments of the primate family, it is fair to assume that these behaviors would also have been found in our ancestors going back to the Late Miocene, about 8 mya or so ago. Our model for early hominins, then, will be drawn from Old World monkeys, especially the macaques, and apes. Of the apes, the model is drawn from both kinds of chimpanzees: the common chimp and the bonobo (the formerly so-called pygmy chimp). This chapter will begin with an examination of the primate social group.

The Role of the Social Group in Primate Evolution

Primates, as a group, are acknowledged as being flexible organisms, able to adapt to local exigencies—environmental and social—not only through the normal processes of genetic evolution but, importantly, *through cultural processes devised in group living.* This latter system, as Hans Kummer noted (1971), is faster than physical evolution, although it requires flexible systems of communication to ensure the transmission of the social artifact. This adaptability, fostered by living within the context of a social group, and accompanied by a big brain with lots of storage space, aided by a system of recall and association of disparate memories, has been the hallmark of primate-to-human evolution.

What was life about in those ancient times? What is it about for contemporary nonhuman primates? Is it a fierce battle to find food, to find mates, to keep competitors from winning? (That is one view; my understanding, however, differs from this. When I see monkeys watching a sunset, I cannot know whether their eyes are merely directed at the light, or whether they are "harmonizing with the ethereal vision.") Certainly, gathering food is a paramount concern, and knowledge about food sources, memories of where these are located, and concerns about what predators might be in the vicinity take priority. However, the tangle that is the *social network* is where the energy goes. What one can and cannot do, when, and to whom are serious lessons to learn. In nonhuman primate societies, the size of a group relates to the resources of the environment, especially sleeping sites and food, yet a society is more than just an aggregate of resource-using individuals. It is characterized by being bound in space, where members of the group are integrated, relating to and related to each other more than to individuals beyond that demarcation. A society is boundless in time as the social group continues beyond the life of any one member so that it is characterized by having generations, or, at the very least, breeders and young (Kummer 1968). Nonhuman primate societies exceed that minimum, having many generations. This means that perforce of there being infants, juveniles, subadults, adults, and old adults, there must be rules regulating conduct between all those individuals.

The old view of the male as being central to the group with females hanging around him, their young attached to the mothers, does not hold true for most Old World monkeys. Female families in macaques form a hierarchy, both within the family unit and between groups in the local group. The oldest female is the focus of the hierarchy, and her children (both genders) tend to have enhanced status, whether through "inherited status" or preferential feeding and nurturing throughout development. Her children form a kind of nobility, born with a silver spoon. The society is structured according to relationships to her, both laterally (sibs) and vertically (offspring). In captive macaques, the status of the mother is conferred on her youngest daughter, who rises to a position just beneath hers. This structured model has not been confirmed for wild macaques (Nakamichi, et al. 2005). Male monkeys born to a high-ranking mother tend to be high ranking themselves. However, males tend to leave their birth group when they mature, and, when entering a new group, must make it on their own merit. Clearly, inherited rank is not helpful here.

Hierarchy in some societies is strict, but in others is much less so. So-called *dominance* refers to differential access to a limited resource. The desired object is yielded to an individual according to that individual's rank in the society. Some nonhuman primate societies are much more casual about relations within the group, and this easy flow is a function of the small scale of the group, where everyone knows everyone else. Where the group is large, an infant begins to know the world from the vantage point of its mother's lap. Its reactions to other individuals are therefore guided by its mother's reactions to those individuals. Initially, all adult females, being large and furry, would be categorized as friendly because mother would not permit a nonfriend or hostile relative to approach. Similarly, any group member allowed to approach, to touch the baby, or groom the mother would be categorized as friendly. Extrapolating traits from these known individuals and applying them to others that look similar produces categories of "friendly," "known," "mother's relations," and the like. Once the little one leaves its mother's body as an older infant and then ventures out as a juvenile, it has as the basis for its behavior the knowledge it has thus acquired from mother's lap. In this way it expands its experience and extends its categories to accommodate the new information. A large furry female that threatens the young animal divides the original category into friends and nonfriends; large, unrelated males who make friendly gestures and offer carriage or a grooming would create a new category of "older friends."

In old age, reproduction slows down, but the social use and need for older individuals remains. The cessation of the ability to breed in nonhuman primates tends to increase with age (Caro et al. 1995), as does the interbirth interval, the time between one birth and another (Burton and Sawchuk 1974). As females age, they trade off breeding with *grandmothering*. The term refers to the babysitting activities of the older female toward the young. The value of babysitting, of a mother leaving her infant with another female while she goes off to feed or rest, has been documented for a variety of Old World monkeys. Recent studies suggest that its origin may be in the evolutionary benefit that comes with the cessation of breeding: the reduction of conflict between generations of females. There is evidence that in natural fertility populations among humans—those which have not used contraception—the average age for having the first baby is 19 years, while the last baby comes at about 38 years (Cant and Johnstone 2008). The coincidence of the interval seems to these researchers to support their hypothesis. Grandmothering assures that the infant is secure, has yet another source from which to learn, and is exposed to an extended network, among monkeys as well as for humans. The mother gets respite; the "grandmother" gets stimulation. Older females spend time with younger females, aid in babysitting, and remain socially viable as a repository of stored information. The capacity to grandmother increases within the primate line from prosimian to ape (Hawkes et al. 1998). The memories of these older females might function in leading the group to food that fruits only once a year or in predicting a predator in a copse of wood along the trail. Babysitting may also have an evolutionary advantage when a female of one species babysits for a mother from a different species. In Kowloon, old Thibetan macaques babysat for longtail macaque mothers, allowing them time off. The behavioral repertoire of the Thibetan macaque was thereby conveyed to the infant who came to know her behaviors, her vocalizations, and gestural communication. This enhanced cross-species hybridization, as the young grew up learning more than one mating "language" (Burton and Chan 1987, 1996). This finding is particularly salient in light of the new suggestion that the human line and the chimp line hybridized over a period of 4 million years before they finally split (Patterson et al. 2006), the flow of genes facilitated by systems of communication held in common.

A society is characterized by shared needs and by its members exhibiting cooperation based on their ability to communicate with one another. Most importantly, perhaps, *a society is an information unit*, composed of all the knowledge that has been stored by its membership. Storage is both genetic, in the DNA of the individuals, and social, in the memories of the individuals (Burton 1984). Paradoxically, both systems are conservative and radical at the same time. Mutations in DNA or chromosomes introduce new material to a genome, but replication of chromosomes (mitosis) is a conservative action, duplicating prior information. The cells that will create the new individual are formed, and the chromosomes in those gametes are allocated as ovaries or sperm, in a process called meiosis. In this process a recombination of genetic material occurs as a matter of course: information from the father's side and the mother's side is passed down in combination to the new being. New combinations of information mean the introduction of new material as a function of the system itself.

In social behavior, the old tend to conserve patterns of behavior, but the young *innovate* constantly through their play. Young ones do silly things, testing each other and their own bodies, exploring the environment and themselves. I have seen young male macaques cover their eyes as they run down the road on three limbs. I have also seen them fling themselves from the tops of trees into deep water, roll over cliffs, "teasing" themselves and grabbing on just before they fall, and even poking and prodding creatures that they perhaps should have been leaving alone.

Innovation occurs in other ways, to be sure, and forms the basis for the development of traditions. The point here, however, is that, both biologically and socially, innovation and conservation of information are constants. This process of give and take is due, at least partially, to the fact that accepting new concepts is hindered by well-established habits. Over time, the information would alter and become depleted or degraded in a closed society. But a monkey society is "semipermeable" (Struhaker 1975), actively filtering who comes in and who leaves, and information is constantly renewed because the young produce new information, or combinations of old information, through play.

Exercising play patterns fixes them within the repertoire of the innovator. The spread of the innovation will depend on the observability of its carrier. With maturation as the only factor involved, the innovation will have the opportunity to move up a narrowing pyramid as the number of animals within the society decreases. With fewer performers these actions are more visible and more likely to be imitated, especially by the young, who don't see the action as "new." Hence, by the time these young are old, the information will have become typical to that group or at least to quite a few of them. If the innovator, during the course of his or her maturation, is a prominent individual, the likelihood of the action being repeated increases (Burton 1984). The information that is stored within a group, therefore, has a natural turnover and progression—it is constantly being replenished (Burton 1972b).

Will this information that is stored within a group be the same for every group comprising a species? To the contrary, field workers know that each local group of nonhuman primates is unique because of certain habits or particular ways of doing things. The term we have given to this differentiating collection of group-shared information is *tradition*. In the past, this term was more readily accepted than the term *culture* for discussing nonhuman primate societies, as it avoided the issue of equality between apes and humans. The notion that nonhuman primates have culture is no longer as controversial as it once was in the scientific world. The term incorporates the notion that while the individual draws once from the gene pool to *be*, it continuously draws from society in order to *become*. In observing contemporary nonhuman primates, we can point to any number of ways in which individuals learn to survive and to get about in their world by means of the lessons learned in particular groups. Although we have already touched upon some of these ways informally throughout the book, it is worth repeating that personality, play, innovation, prominence, dissemination, and fixation or loss of the pattern are several of the routes by which traditions come to differentiate one group in a species from another.

Let us look, for example, at all the information that must be processed in regard to food. The basic information is "what to eat." This information is transmitted to the monkey infant through smell or the taste of the food it takes from its mother's mouth. It is apparently also gained from tasting *in utero*. When the mother removes unsuitable food from the infant's mouth or pushes its little hand away from tempting but perhaps toxic plants, the infant comes to formulate categories of "edible" and "inedible." As the infant matures, the category expands, with new experiences gained as it leaves its mother for longer and longer periods and associates with its peer group. The young animal observes, models (Hall 1968), and taste-tests. The growing youngster must also learn the "conformity" concerning with whom it may eat and under what circumstances—a set of information that is quite taxing as it varies greatly from situation to situation. The contingencies involved might read: if adult female 1, then eat; if adult female 2, then run unless female 2 is resting, with an adult male, or being groomed, and so on. Bridget is always friendly and ready to cuddle. Wilma threatens but doesn't do anything. Joan is nasty, and while Mark is gentle with youngsters, Ben is gruff. Knowing how to get along depends on remembering who is who and how that individual is apt to behave. It is apparent to field observers that this kind of information is constantly being processed. These traditions are socially endowed and derived. The quantity of information processed depends on the individual's ability to store as well as to generate it and on there being individuals known to each other.

Traditions are group specific. The pool of group information available depends upon the number of contributing individuals within the group. The information reposing in twenty-five individuals is greater than that in only five, but also depends upon the age breakdown of the group as each individual brings his or her personal experience into the group pool. Furthermore, each individual absorbs only a portion of all patterns available in the group. Age and status are major factors molding performance of traditions as well as absorption of them. The older animals are the repository of wisdom, meaning the amount of information as well as the implementation of it. They function as the library in a nonliterate world. The prominent position that older animals hold in most monkey societies makes them central to the group—literally, "seeable." The actions that older monkeys display, therefore, filter through the group. Such actions tend to be conservative, the repetition of patterns that have been witnessed before.

On the other hand, each individual has its own distinctive personality, the result of the particular expression of its genes within the unique environment of its mother and later its peer group. Students used to balk when I called a monkey a "person." The concept that a nonhuman animal could be a *person* was not acceptable. Because the sum of prenatal and developmental influences enables the individual to self-form, a being is a person. Its responses to light, heat, other individuals, and even itself are based in its physiology and, through repetition and testing with others in the group, those responses grow, become more complex, and continue to express what becomes its "self." Individuals innovate. By chance, play, or design, new patterns are expressed that others may adopt (Hall 1963; Burton and Bick 1972).

Traditions, then, are carried by *individuals* from one group to another when they leave their natal group as adults. In some species it is females who leave (for example, chimpanzees); in others it is males (for example macaques and baboons). Regardless of gender, it is the performance of a behavior and its witness that enables it to be remembered and enacted at another, appropriate time. This *sampling* of the inventory of a culture depends on social organization. The fission-fusion pattern of dispersion and regrouping, in a kind of ameboid movement, is typical, for example, of the hybrid macaques of Kowloon. It means that segments containing upward of sixty to eighty individuals can be visible at any given moment, but that each segment will flow back into the main body of the group, and different members will flow into a newly forming "pod." The casual observer might mistake this sample as a stable group, rather than a dynamic source of information transfer. The fact that not all local groups share the same traditions indicates that innovation, like mutation, is a local phenomenon that spreads with dissemination.

FIVE

Whether or not traditions are adopted may be influenced by the position the innovator holds in the society. Hence, an entire local group of a species may hold to a tradition not shared by other groups of the same species. In Gibraltar, I found that the adult male monkeys held and nurtured babies more than did males in Morocco, the original home of the Gibraltar monkeys, and, surprisingly, they did this nurturing even more than the related neighboring group a quarter of a mile away. The adoption of the pattern, as I saw it, was largely a function of the then-leader male, known as Mark, who, in effect, *defined* malehood and leadership by his interpretation of how males take care of babies. Youngsters growing up in this group observed what he did and how he did it. The pattern then became part of their behavioral repertoire, and the exercise of this behavior awaited only maturity and opportunity.

Traditions can be transient. Individuals change, circumstances alter, the environment is modified, and so the context changes and the traditions may fade. Male care changed over time in Gibraltar; its expression became less intense as the population size grew, and recently a new tradition has developed. Alison Carroll tells me that she has witnessed the Gibralter monkeys clipping a leaf and holding it between their lips as an apparent signal that grooming is about to begin. Baboons studied by Shirley Strum in Kenya not only created traditions of hunting, but forgot them as well. Males began hunting small prey—dik-diks, hares, and the like—but after only half a generation or so, the tradition faded (Strum 1981). Ironically, it is precisely this process of innovation, adoption, practice, and loss that marks the birth of traditional or cultural patterns. The transience of tradition actually affirms the process.

How pervasive a tradition is depends on a variety of complex factors. Role behavior defined by traditions in the group, for example, can last through several generations. What I found in a survey of twenty-one monkey societies was that even the most biological of behaviors—food getting or baby rearing, say—had a societal definition. Allowing individuals to assume leadership roles—warning, protecting, and moving the troop—depended on the group's acknowledgment that the actor in these cases had sufficient *knowledge* to warrant the group's trust. Furthermore, assignment of gender to a task seemed arbitrary over all the societies, except, of course for bearing and feeding infants. Monkeys perceive the gender of neonates visually and through scent and taste as they greet and lick the newborn. It is a rare event to see a monkey baby being born, although I have been close, coming upon a mother just minutes after her baby's birth. As the baby emerges, it crawls up the mother's tummy and comes to nestle in her arms. The afterbirth dangles from the cord, which is left to wizen and finally fall away. Sometimes the afterbirth is attached for a while and can get caught in branches, resulting in the infant being herniated. Perhaps it is the smell of the infant that attracts others, or perhaps it is anticipation of the birth, but soon after the baby is born, the others come to groom the new mother and her baby. The mother looks tired—birthing is not a simple task anywhere in the primate world.

Male involvement with the socialization of the young may be *passive*, where the male is merely the information source of some of the traditions of the troop—for example, what to fear, where to forage, when to give the alarm bark, what is edible. Alternatively, the role may be *active*, ranging from carrying the young, as in some leaf-eating monkeys and some baboons, to the even more intense involvement seen in groups of macaques. In Gibraltar, at one time, only young males were allowed to carry babies, taking the infant from the leader male macaque who was responsible for the infant from its birth, carrying it, encouraging it to come toward him (to walk?) and to "natter." This social gesture was developed from the infant's lip-smacking. When the baby did so away from the breast, both parents enthusiastically responded with their own lip-smacks until, after a couple of weeks, the infant was doing this as a social gesture, a greeting (Burton 1972b).

The social task of leading the group requires that the leader know where to go and when and that the others acknowledge this by following. The leader—and it can be of either sex—must be familiar with the routes to feeding and sleeping areas and be able to predict where contact with other animals may occur. In addition, the leader must know how to mobilize the troop, which is a function of being able to indicate not only movement itself, but direction—roles called "initiater" and "determiner" (Kummer 1968). The actor or role player is followed, and in this context that means chosen, perhaps on the basis of position or status, but certainly on what amounts to troop presumption of an adequate store of knowledge on the part of the leading animal. In the Barbary ape of

Gibraltar, for example, an older female, knowing the terrain, would stand up where all could see her and wait. Slowly the group would form behind her. The male's greater participation in troop movement as determiner is partially related to his role in protecting the troop and partially a function of observer bias: a larger animal is simply more prominent.

Monkeys are aware of themselves; that is, they "know" they are organic forms bound in space and different from others in their group. Turkeys may hear others gobbling and *continue* gobbling themselves as they are part of the collective, but monkeys quickly develop a sense of their separateness from others. It has been mentioned that the sex of the newborn is registered by others who come to greet it and to groom the mother. The cues are smell, sight, and taste as the new baby is fondled, licked, and cleaned. This registration of sex is influential in subsequent contact with that infant. Psychology experiments have shown that mother monkeys act differently toward the infant according to its sex, thus setting patterns of reception of gender-based behaviors (Kummer 1968). In addition, the health status of the mother will affect her infant. A healthy, well-positioned female will enable her offspring to feed well, both directly through her own milk and indirectly by providing good access to preferred foods. The quality of mothering also makes a substantial difference to the maturing youngster. Goodall documented how, in chimps, older mothers tended to be more laissez-faire than younger ones (1968, 1986). In monkeys, as in humans, there are indulgent and permissive mothers and those that are inconsistent and irascible. The difference in the offspring is obvious: the young of the former are calm and confident, while the young of the latter are nervous and not as ready to cope with others. A well-nurtured offspring gains the self-confidence necessary to take a position within the troop; conversely, a monkey protected by a high-ranking mother can lose it all depending on its childhood behaviors. Sweetpea was a hybrid macaque juvenile in Kowloon. Sweetpea's mother was huge compared to the other monkeys and defended her no matter how much she teased, pulled hair, bit, or chased. All Sweetpea had to do was to hide behind her mother—nobody messed with mom. Times changed. The young female adult was now unprotected, and her peers had not forgotten. Instead of the confident, aggressive youngster, she became submissive, retreating from those she had formerly bullied.

Identity of the individual is not conferred, but occurs during development. The process, being dynamic, continues throughout the life of the individual. While detailed observations of what a monkey or ape experiences can be made, observations of what it has witnessed are not as easy to see. Given the long road to adulthood, what the infant absorbed from childhood experiences may become reflected in its adult performance but cannot easily be identified by an observer. I have been fortunate to be able to revisit one site for a number of years, so I have been able to "catch" monkeys observing other monkeys doing something and then, years later, see the observer animal repeat the action. This is why from the outset I considered that behavior could not easily be relegated to gene action without considering all of the environments to which the animal had been exposed since birth. To complicate matters still further, the infant soon leaves its mother for a peer group. Not only will it absorb information from this source, but also from the older youngsters that play or intervene with the peer group. Contact with the next older age group also occurs in a variety of other contexts, such as grooming, sleeping, huddling, feeding, and troop movement. Whatever the activity, there is a lot of variation in behavior patterns presented to an individual.

As the infant moves to the juvenile stage, it leaves its passive role of troop cohesion for an active role in protection, particularly in giving warning barks in response to strange stimuli, like birds—some species of which prey on monkeys. Eager youngsters bark at any bird they see, but the older animals don't always react. The youngster soon learns to discriminate between types of birds according to the response to its barks. The burgeoning understanding of social responsibility is accompanied by an apparent new perception of the group, such that the class of "older individuals" is broken down into male and female, and then into particular individuals. At this point, the decision of whom to emulate becomes further complicated by the youngster's perceptions and goals.

That nonhuman primates—at least chimps—are political animals has been well documented (Munckenbeck-Fragaszy and Mitchell 1974), and it is likely that they maneuver in society according to some desire for the attributes of power (de Waal 2007). I have seen political jockeying in monkeys, too. Monkeys manipulate other members of the group by doling out or withholding information; they can deceive, are politically acute, and seem intent on formulating plans for specific ends. Grooming, for example, is used manipulatively by monkeys for a desired outcome. It is the social adhesive cementing relationships within the group (Burton 1972b). When Charlie grooms Alice, he is attempting to forge a bond, form a coalition, smooth the way to mating; conversely, not grooming an individual who presents a part of its body for that pleasure clearly represents rejection. Grooming is the only "currency" of nonhuman primates (Gumert 2007) and is distributed or withheld according to complicated and little-understood variables. The existence of this currency would have been a strange thought 20 years ago, but has now become accepted by scientists under the rubric of "market behavior."

One spectacular instance of the use of grooming for a political end concerned an adult female, Joan, who at the time was 9 years of age and the youngest of the fully adult females. She was observed over 3 years in a complicated quest for power. The maneuver involved Bridget and Wilma, two older adult sisters, and Pat, Bridget's daughter. Initially, Joan isolated Wilma by consistently threatening, chasing, and generally rebuffing her, especially when approached by Bridget. The following year, by consistently grooming, sitting with, and generally remaining near Pat, Joan was able to keep Pat from Bridget. In this way the kin group of Pat, Bridget, and Wilma was broken up. When friction arose within this kindred, Joan allied with Pat against Pat's mother, Bridget. This coalition enhanced Joan's position against the older females. However, this power play must not have been sufficient for Joan since she began consorting with two young males and with them began wandering away from her natal group, as if to establish a new group elsewhere. In so doing, however, she overstepped the human tolerance for monkeys in Gibraltar, and the wandering monkeys were captured and sent to zoos.

Chimps and Bonobos: The Ape Model

Our discussion of tradition and its transmission has focused largely on the description of society among Old World monkeys, but chimpanzees provide a picture no less complex. Early research into chimp culture by Jane Goodall and Bill McGrew (Harvey et al. 1988) has now been affirmed by many others (Wrangham et al. 1994). Although for years scientists had depended on information coming in about the common chimp, *Pan troglodytes*, to help in reconstructing the ancestral condition, primatologist Adrienne Zihlman, in the 1980s, focused attention on the bonobo chimp, *Pan paniscus* (Zihlman 1996). The addition of the bonobos to the knowledge about chimpanzees expanded the ape model. The differences between the two species are striking, yet they diverged from each other perhaps only a million or so years ago. The bonobo is a smaller animal, with sexual dimorphism little expressed in size, robustness, or teeth. They are more delicate looking than the common chimp, lacking the robust aspect of the common chimp's skeleton and skull. Their skin and fur is black, their ears small. They live in small groups ranging in size from between five and fifteen individuals, with all ages and sexes represented. These smaller groups often aggregate into larger communities of up to sixty members. These larger groupings occur when fruits ripen in a particular area, and all can profit from them.

Nonhuman primates know their world in three dimensions. Sight, sound, and scent operate to inform them of the location of others, so a "group" may well extend beyond the visible animals a field worker encounters. Trees do not constitute an impediment to the chimps knowing who is where. The chimpanzee groupings, therefore, are momentary manifestations whose members belong to a larger aggregate. The fission-fusion pattern with individuals coming together and then parting is found in other kinds of nonhuman primates as well. The community is the basic unit for bonobos, with the smaller units being the foraging units (Badrian et al. 1981; Kano 1983). Unlike the common chimp, all-male groups and nursery groups of females with young are seldom seen. Rather, the bonobos live in mixed groups, where the sex ratio approaches one male to one female. While it is females that leave the natal group, they tend to stay with the larger community. Social behavior in these chimps differs from common chimps as well. Bonobos are characterized as peaceful, with aggressive interactions at feeding sites amounting—at least at one site in Zaire—to only one incident in 68 hours of observation (Badrian and Malenky 1984). They are highly social, using sexual behavior and gestures to maintain social cohesion, and are characterized as "democratic" in that males do not dominate females, and females can readily take food from them, a

FIVE

behavior not seen in common chimps where a good deal of begging, cajoling, and personal history precedes any donation to another.

As with most nonhuman primates, positive affiliation is reinforced by bouts of grooming. Unlike the common chimp, where grooming is frequently between members of the same sex, bonobos groom the opposite sex a considerable portion of the time. Their use of sex, however, is unique among nonhuman primates, and rivals ours! Sexual contact can be heterosexual or homosexual; young and old can also form a consenting pair (Kano 1983). Genito-genital rubbing, as it has been termed, is used to reduce tension, for example between females during episodes of food sharing. It should be noted that food sharing or object sharing used to be considered a distinctly human activity. It indicates the recognition of another being, its wants, and rights. There are gradations of food sharing. Permitting an infant, relative, or friend to take food is one level; proffering food with the open hand or by presenting it is quite another as it implies the will to transfer an important object to another individual, recognizing one's own need or desire and imputing it to that other. Where the exchange of favors accompanies food sharing, the sophistication of the use of contact is particularly striking.

The degree of tolerance in a society of bonobos distinguishes them from common chimps as well. Adult females in the common chimp are tentative in their approaches to adult males, especially if the latter are hyped because a rainstorm is coming or there is newly killed meat. Not so among the bonobos. Among them, quiet sharing of plant food (Kano 1983) or even fresh animal kills, like small hoofed animals, have been observed (Surbecka and Hohmann 2008). In this latter case, copulation precedes sharing (Kuroda 1984). This latter behavior is known even in New World monkeys—for example in the saki, *Pithecia pithecia*. Food getting may be a priority in chimps, but fully two-thirds of their active time is spent in other pursuits (Badrian and Malenky 1984).

Common chimps are quite different from bonobos. Chimps are usually found in small fission-fusion units. Sometimes the unit is quite small, between two and four individuals, and compositions vary. A unit can be a nursery group, made up of females and children; sometimes it is an aggregate of several large families; sometimes a mating group, with male, female, and her young; and less frequently a mixed group of all ages and both sexes. Goodall estimated that 30 percent of her chimp groups at Gombe were mixed (Goodall 1968), and that number goes up to 47 percent at Mahale (Nishida 1979). Large groups containing only males patrol their territory, seeking and destroying chimps from other groups (Nishida 1979). Footage of this behavior is terrifying: the hair on the neck is erect, the eyes are glazed and focused, the faces tensed in determination. This seek-and-destroy behavior began about 20 years ago. That chimps are capable of it is part of their makeup, a physiological and motor response based on other behaviors. That they enacted it may very well be due to environmental degradation and encroachment by humans.

Male common chimps roam over a widespread area, which corresponds to the areas in which females, in their family and nursery groups, range. Males are most attracted to females in estrous, with whom they spend the most time. Each sex has its own life strategy. Males are considered to maximize reproductive opportunity, while females emphasize foraging benefit. Food getting takes nearly half the active part of the common chimp's day, and, in comparison to bonobos, they go further to get it and get less of it from the trees (Goodall 1986). Common chimps also share food from both plants and hunted meat. There tends to be a division of labor in that it is the males that more frequently hunt and eat meat, while the females are more apt to hunt termites, manufacturing the tools that enable them to do so.

Higher-order Capacities in Nonhuman Primates as a Guide to the Ancestor

We have noted that information is stored within social groups of nonhuman primates by means of traditions, which are socially endowed and derived and which serve to differentiate various groups from one another. Furthermore, the quantity of information being processed depends on the abilities of individuals to store and generate it (in addition to the requirement that there be individuals known to one another in order to share the traditions). This suggests that the ability to perceive and absorb societal patterns depends on higher-order processes of the individual group members. Let us turn to a consideration of these processes and what they can tell us about the likely capacities of the Ancestor at the time of the first use of fire.

The formation of group traditions is characteristic of monkeys, as we have seen. Apes do this and even more as their cognitive abilities exceed the capacity of monkey brains. What sorts of cognitive processing can we observe and infer from the behaviors of nonhuman primates in helping us reconstruct the likely abilities of the Ancestor of millions of years ago? Consider first the higher-order capacity of *classification*. Because of the numbers and types of pieces of information that reach the primate brain, the number of decisions that can be made is a function of the amount of specification that is possible. To assist in this decision making, classification of information is adaptive, permitting sorting of the abundance of data coming in. What underlies this ability is plausible, inferential reasoning based on recognized similarities between things. The more factors used to classify objects or actions, the more precise that classification can be. Furthermore, it requires no "words." The child psychologist Jean Piaget, on the basis of his observations, concluded that the first type of thinking evidenced by the human child occurs outside of language. The basis of logical thinking, it seemed to Piaget, derives from actions, and from the knowledge the child abstracts from the objects upon which he or she acts. Such action occurs before expressive language and before language-mediated thought.

A forest monkey, for example, must be able to process data regarding trees. The animal's ability to deal with the requirements of forest life is enhanced by being able to recognize trees with certain kinds of branches and able to hold the animal's weight, trees holding food, or trees where snakes, birds, or big cats can hide, and so on. Each of these single constructs enables the animal to categorize the item by function or attribute. Field observers note this process of analysis when they record head movement, eye position, movement of eyebrows, and actions with reference to objects.

Classification, as an act, includes several other kinds of abilities. Primary classification for a monkey includes identification of objects through smell, touch, and sight, all of which allow for the creation of images. The differences in the images permit the monkey to discriminate between them, and they become the basis, or data, upon which classes are formed. These classes, by their inclusions and exclusions, permit the animal to predict outcomes. By so doing, the monkey is constructing a type of taxonomy. The ability to categorize in this way is suggested by observed examples concerning food behavior. In the first example, monkeys transplanted from one habitat to another are able to find food relatively easily in a similar habitat. But by what means do they apply what they know to the new resources? Is it the shape of the plant, its odor, the nature of the leaves, or just trial and error? Plant identification has not always been successful. Japanese monkeys transplanted to Texas died from eating a local weed. However, relatively few of them died before avoidance of the plant became typical. Assumedly, the class of "inedible" was increased by the features of this toxic weed.

Secondly, it is no less true that classification and flexibility are required for the *foraging behavior* of nonhuman primates. Different species react in different ways to how much food there is, as well as what the quality of the food may be (van Schaik et al. 2005). Flexibility is the hallmark of nonhuman primates. Primates show this flexibility not only in the choices of food they make, but in the apparent ease with which they adapt to varying circumstances despite real hardship. Under harsh conditions, primates will seek (as indeed humans do) what are called fallback foods. These might include leaves, pith, bark, vegetation on the ground, certain insects, and staple fruit such as figs that are not as high in fats or carbohydrates as are preferred foods (Knott 2005). Environments change, and sometimes dramatically so, and primates have an impressive (if intuitively obvious) repertoire of methods to deal with seasonal changes in food availability of the preferred foods: they can move to a location where that food may be available; they alter their diets to include less preferred items; and they change their behavior to coincide with reduction in nutrients and the consequent reduction in energy (Brockman 2005). Fallback foods are currently considered to have had a strong evolutionary effect on the jaws and teeth of orangutans and australopiths, whose thick molar enamel enabled them to eat tough and gritty foods (Vogel et al. 2008). The cues that enable nonhuman primates to find preferred foods can be quite subtle. The Mediterranean climate in Gibraltar means seasonality in the growth of prickly pears. Whether the monkeys—and it is usually an older female—are cued by scent drifting over a considerable distance, time of year, or moisture in the air, I do not know. But with an uncanny regularity, they find their way to that delectable fruit once a year.

Not only are monkeys able to form taxonomies, they are also able to prioritize in the presence of several competing events or stimuli. This extension of classifying, of making taxonomies, is observable in the choice of response pattern based on known (previously observed) options. When confronted by an attacking adult female, a screaming infant, and a delicacy, a monkey can be seen to freeze, look around, and then act. The choice of action—to flee the attack, retrieve the infant, or grab the delicacy—cannot be simply predicted by the observer on the basis of which choice seems best to correspond to "survival." The choice of option is predicated on the personality, gender, age, experience, predictions, and other factors unique to the actor (Burton, 1984). Sometimes choices are difficult to make. A film on baboons by Shirley Strum records an adult male choosing to remain with an infant he is babysitting (perhaps with an eye to eventually mating with its mum) rather than go with a female whose alluring sexual presentation he clearly wants to follow (Strum 1998).

Forming hierarchies also implies putting things in sequence according to a given set of features—that is, ordering a series. Monkeys require this ability in order to "solve" ordinary problems. If an animal were to come suddenly upon an abundance of food, for example, it would have to organize its tactics of retrieval—and this in the presence of others to whom this plenty is also appealing. There is only so much a monkey can carry—even with cheek pouches, an attribute unique to Old World monkeys. Therefore, through a sequence of gathering food, filling pouches, leaving the scene, and returning for more, a serialization of food-getting behavior occurs. When the sequence is time bound, it suggests the primitive formation of chronologies. Unlike apes, monkeys apparently live without knowledge of the past. They may plan for an eventuality—a future—but reaction to a memory is like a stimulus newly received. (I have seen monkeys sitting quietly, doing much of nothing, suddenly leap up and run. What memory triggered the action? What image was conjured that stimulated the animal to react?)

What of the role of *toolmaking* in understanding primate cognitive abilities? Toolmaking has now been amply documented, and there are numerous videos illustrating this behavior. Goodall listed sixteen societies of wild chimpanzees that are known to make tools (Susman 1987), and McGrew has raised that number to thirty-two, including societies of released as well as wild chimps (McGrew 1992). Habitual patterns of tool use include using a tool to obtain food, the classic example being the termite probe, which is fashioned, carried to a remembered site, and then inserted into holes in the termite mound to extract the insects. Leaves are used as sponges or napkins and as grooming instruments. Hammers are made to bash nuts, and observers have even noted how mothers teach offspring the proper techniques by placing the template to be worked into the right position while the juvenile attempts to reproduce the mother's working pattern (Boesch and Boesch 1990; McGrew 1994). Since stone is scarce in the Tai Forest of the Ivory Coast, chimps remember the location of particularly useful stones. They can readily lift a 9-kg (20-lb) stone to smash a nut or apply delicate taps. Chimps use tools to dig into the ground to retrieve storage organs of plants. Monkeys do this as well but can use only their hands and cannot penetrate the earth deeply without digging tools. In dry habitats of Senegal, a variety of woody or viny-supple vegetation is used by chimps (McGrew et al. 2005), for example, but in Congo a more sophisticated set of implements, specific to whatever task is to be undertaken, has recently been found. There is a puncturing stick with which a hole is excavated into the termite mound. Then a twig, carefully and skillfully shredded at its tip to make a brushlike object, is inserted into the mound, its complicated surface allowing more termites to cling and be extracted (McGrew et al. 2005). The kit is specific to *place* and to this task, and what is more, the source of the tools is chosen for their attributes: the puncturing stick comes from a particular tree, but the brush-tipped probing tool comes from a species of herb. The fact that the chimps seek tools from particular sources and then fashion each tool differently and according to purpose speaks volumes.

Since the late 1980s, chimps have been observed in a variety of locations in Central and Eastern Africa using tools to retrieve honey (Yamagiwa et al. 1988). One paper describes how sticks found at the base of active bee colonies appeared to have been fashioned specifically for the task of retrieving honey. These flexible twigs from vines and lianas were stripped of leaves and the bark from both ends had been removed.

The sticks were peeled as well as chewed, and the frayed ends smelled of honey and bee brood, affirming that they had indeed been used for the purpose of gathering from the honeybees (Brewer and McGrew 1990). Another study describes how different tools are used for different types of honey gathering (Stanford et al. 2000), and there are similar observations of chimps using different tool sets to get termites out of their mounds (Bermejo and Illera 1999; Sanz et al. 2004). Other animals in the vicinity, such as baboons, civets, and gorillas, destroy bees' nests but do not use tools (Kajobe and Roubik 2006).

In the Moto community of Central Africa, the chimps use two different kinds of wooden tools to procure a meal of termites. Termite homes are either mounds above the ground or subterranean chambers. Chimps use specific tools for each kind, depending on its structure. They use one set of tools, including larger sticks, to puncture the subterranean mounds and break through the underground chambers and another set to perforate above-ground mounds and extract the insects. A *set* of tools in the chimp context refers to at least two components used one after the other to achieve the same goal (Brewer and McGrew 1990). The larger puncturing sticks are often left at the termite site and used by other visitors; the thinner sticks, however, are deliberately manufactured, often a distance away, from woods especially chosen for their useful properties. The leaves are removed by mouth or hand, and the end frayed into a brush. Before insertion into a hole, the frayed end is moistened, rather like the procedure involved in getting thread through the eye of a needle: moisten, twist, insert. The disturbed termites gather by the hundreds to expel the intruders and hang onto the frayed brushlike end. They are extracted and eaten. They are removed from the twig either directly with the mouth or by sliding the twig through a fist and gathering the termites at the top (Sanz et al. 2004).

The observers of this chimp behavior used video cameras controlled remotely to capture these actions. Although it took a little while for the chimps to accept this new object in their termite excavations, they went on about their business. There are several important points about the observation: by using cameras to observe them (the chimps quickly ignored the sound), chimps were doing their thing without relating to people at all; secondly, the chimps were using sets of tools fashioned for a specific purpose, from specially chosen material, and manufactured, at least in part, at a distance from the location in which they were used. This indicates a high level of memory, purpose, foresight, and planning. In addition, as the researchers note, and in some respects the most important aspect, the tools used by the chimps would leave no archaeological trace. The significance of this is enormous as it reflects on the Ancestor, a creature diverging from the apes and who we can assume had at least the same capabilities.

What monkeys and chimps *cannot* do, however, also indicates the path hominization took, and that path is connected, to some degree, with the more abstract capacity to take the point of view of another with whom one is sharing an experience. While chimps may have a nascent theory of mind, for example, experiments have shown that when gazing in the same direction as someone else, they may not be paying attention to the same thing within that gaze (Tomasello et al. 2003). Perhaps not, but I once shared a gaze with an old female monkey in Gibraltar. One time, when carrots were distributed to the monkeys, knowing she might not get her fair share, I looked at a location that other group members could not see. She followed my gaze to it, and slowly made her way down. When I was sure no one else had noticed, I followed her to the rendezvous and presented her with the carrots. This intention-understanding ability would suggest that the behavior has ancient—if undeveloped—roots. While a *gaze* is understood by monkeys (Deaner and Platt 2003; Goossens et al. 2008), monkeys do not *point*. Chimps do—an important intellectual behavior that presumes a higher level of symbolic referencing.

Pointing is redundant to the gaze: you point in the direction in which you are looking, and chimps are known to do both. Evidence shows that a chimp will alter its gesture to take into account the gaze of the colleague (Povinelli et al. 2003). In hunting colobus monkeys, chimpanzees *coordinate* their movements using their eyes: gazing in the direction they want another chimp to move to prepare to capture the monkey. The redundancy between pointing and the gaze makes the action more clear as the viewer understands *exactly* where the actor wants the viewer to look. However, chimps, compared to young human children, apparently will not understand that another chimp is focused on something else within the same gaze direction; they seem not to be able to imagine how things may look when those things are seen from a different viewpoint; nor do they seem reliably to grasp prior intentions (Tomasello et al. 2003). There is also the possibility of misaligning a cause with an effect. Superstition is based on this incorrect juxtaposition of events. For example, in the wild, the loud bark made by some kinds of forest monkey is usually given by an adult male toward evening as a call to coalesce the group and to inform them of the direction in which he is about to move. In a cage situation, however, the call becomes associated with something else—perhaps feeding time or lights on. Its meaning and purpose thus change. In any case, the Ancestor will not have had such capacities as understanding the viewpoint or the beliefs of others (Tomasello et al. 2003). Some would argue that such a highly abstract ability requires speech. And it is to this issue, of speech and its relationship to culture, that we will turn next.

Do Nonhuman Primates Have Culture?

Perhaps the most controversial issue of all concerning the relationship and continuity of lineage between humans and nonhuman primates is whether nonhuman primates can be said to exhibit *culture*. An exhaustive evaluation of whether or not culture exists in nonhuman primates was published in the late 1990s by Bill McGrew. It was based on the criteria proposed by the famous anthropologist, and one of the fathers of the field, Alfred Kroeber. In the late 1920s, Kroeber defined what culture should look like (1928, 331, in McGrew 1998):

1. A new pattern of behavior being invented, or an existing one being modified
2. Transmission of this pattern from the innovator to another
3. The form of the pattern being consistent within and across performers, perhaps even recognizably stylized
4. The pattern persisting in the repertoire of the acquirer long after the demonstrator is absent
5. The pattern spreading across social units, be those families or clans, or across troops or bands, in a population
6. The pattern enduring across generations

Over decades of primate research, it has become increasingly clear that these criteria are indeed met in the behaviors of nonhuman primates, especially in chimpanzees and bonobos. So why is this issue still controversial? It seems to me that there are both emotional and more scientifically based reasons for lack of a broad scientific consensus on this question.

Ancient epics, like the tale of Gilgamesh, clearly divided the world into humans, as the governors of nature, and animals, as objects requiring governance. The post-Gilgamesh human has not been inclined to share the planet either. Scientists, too, have been reluctant to admit "lesser beings"—in this case, monkeys—into the intellectual sanctuary. But apes, especially chimps, do belong. Evidence of their ability to think, innovate, deceive, manipulate, and engage in other higher-order mental processes have been well documented. I believe that, to some extent, the reluctance to accept that other organisms also depend on the mental products of their behavior stems from the realization that we must therefore recognize that there are other sentient beings and that we must share the planet with them.

Students of animal behavior are required to be careful in assigning human traits to nonhuman beings. Attributing human characteristics to nonhumans is *anthropomorphism*. Kitty is not *really* "telling" you to open the patio door; it is not insight that brought his paw to the door. Rather it happened, as it did to me, because Kitty pushed the dog port—at random—and it clacked. I responded by opening the door, and so he conditioned *me* to go to the door and open it when he taps. But students of animal behavior are also warned not to *deny* the abilities of animals. It took the soul-wrenching experiments of Harry Harlow, who separated infant rhesus monkeys from their "dams" (saying "mother" was a big no-no in reference to animals in the 1950s), to determine that infant animals are attached to (love) their mothers. He showed that the infants preferred soft terrycloth mother substitutes to the wire "mother" with a bottle, even when they were extremely hungry (Harlow 1958). Perhaps not a surprise to many, it was incontestable proof that animals have feelings. The great Aristotle had denied animals a conscious soul, so how could they have feelings? Over the course of human inquiry "soul" had become "mind"—that coextension of the brain responsible for reason, perception, will, and feeling. For the longest time, denying a *soul* to all animals also made it impossible to grant that nonhuman primates could have *mind*.

It took the work of the animal physiologist Donald Griffin (Griffin 1976) to raise the issue of *consciousness* as an aspect of mind in nonhuman animals. Consciousness is a quality of mind and subsumes self-awareness, feeling, and the processing of feeling, knowledge, and the capacity to interpret and organize information about oneself in relation to the environment. Griffin noted that the storage and recall of thousands of environmental images is the hallmark of nonhuman primate mental activity (Griffin 1976, 1984). They need to know not only where they are going, but also where they have been and what happened while they were there. It is a necessary condition for awareness. Griffin argued that because both animals and humans exhibit rapid eye movement (REM) sleep phases, and because REM sleep is the dream phase where the self is projected into action in a created or remembered scenario, an animal's dreams, just like a human's, must reflect the projections of self in a scene, the primitive hint of displacement and recognition of time.

Studies like these dealt a serious blow to negative anthropomorphism—the insistence that we are absolutely different from other animals and thereby cannot presume that what *seems* the same is indeed the same. Behavioral science had been held back for a very long time when this view was the accepted one. New technologies now exist, however, that permit the viewing of intimate details of the brain never seen before; similarities of function show up in these studies. Acceptance of the implications of these observations has allowed the study of the behavior of nonhumans to blossom—and with it an incredible wealth of understanding (e.g., Cheney and Seyfarth 1990).

We have discussed the role of nonbiological transmission of behavior from one generation to another in defining culture. What about the roles of intelligence and of speech in our definition? Some in the scientific community still feel that cognition depends on speech, and therefore culture depends on speech as well. Speech is specific to humans, although all animals have systems of communication. Monkeys and apes use a variety of vocalizations. All facial muscles are used in conveying complex and even ambiguous "statements." In *Macaca sylvanus*, for example, the eyes may show threat, while the mouth makes the social gesture of approach (Zeller 1992), and vocalizations can also be combined meaningfully to create new statements, identifying the speaker, the situation, and direction of travel at the same time (Arnold and Zuberbühler 2008). Thus, the throat, lips, tongue, and nasal passages are all used to produce sounds, the meanings of which are understood in the group. None of these abilities to combine units of meaning into longer, more complex units with additional information were considered previously to be within the province of monkeys. Their confirmation in monkeys suggests a much greater antiquity, and slower progression toward speech, than had been assumed.

Beyond the use of vocalization and facial expression in communication, apes (chimps and bonobos) have been taught to use sign language with their human mentors and, most impressively, have taught it to infant chimps. Sign language is now considered to be a legitimate form of *human* language (Bickerton 1990), and we might then equate chimp use with human use of it. So can we continue to argue that cognition—thinking—cannot occur without speech? The studies of Jean Piaget, mentioned earlier, suggest that the human child exhibits "thinking" long before it can speak. And studies of deaf children (Tomasello 1994) indicate that speech is separate from thought, as deaf children develop spontaneous sign systems with all the properties of language. We can conclude, then, that intelligence exists without language, although language requires intelligence. But is intelligence sufficient to develop culture, if we define culture as consisting of values, deeds, and beliefs—abstractions that precede patterns of behavior passed on from generation to generation? The scientific and popular literature is rife with nature stories about what apes can do within ape-human scenarios or in laboratories, as well as what they do on their own in the wild. So perhaps the issue is not cognition *per se*. Cognition depends on intentionality, where the animal makes choices about what it is doing and even how (Burling 2005). Rather, the issue concerns whether these cognitive abilities are sufficient to form culture as we define it.

Some scientists argue that they are not sufficient. Analyzing the process of social learning, for example, Michael Tomasello has argued that what chimps do is *emulate*, not imitate (van Hoof 1994). They attempt to reproduce the end product of a behavior, but cannot copy the details by which this is achieved. Tomasello and his colleagues (1994, 2003) go on to consider that "faithful reproduction," or uniformity, is essential to culture as is also "history"—that is, an accumulation of modifications of a pattern. Certainly the variants of a pattern are found in the traditions of nonhuman primates. As for uniformity, rote memorization and faithful reproduction seem to me to be the antithesis of adaptation and innovation. However, Tomasello argues that these two latter traits are probably of fairly recent origin, that is, accompanying *Homo sapiens* himself, but certainly long after the divergence of chimps and apes. But human culture does not require insightful understanding before culture is acquired. Why ask more of chimpanzees? If chimps are not to be granted cultural behavior, certainly what they are capable of doing and what they actually do are well nigh close, very close indeed. The parallel to biological evolution was presented earlier where, in the ordinary meiotic process of forming the germ cells, the crossing over of parental chromosomes ensures a new "reading" of genes, and indeed, where a mutation is a misreplication. The debate on what—exactly—*is* emulation or imitation and whether or not apes do it, continues (Byrne 2002).

But let us return to the issue of *speech* and whether the development of speech in the line that eventually became *Homo sapiens* is as abrupt and discontinuous from the behavior of other primates as it has often been argued. Alfred Kroeber, one of the many fathers of anthropology, asked whether chimps do not speak because they do not have the necessary structures or whether it is because they have nothing to say! It is hardly the latter. The anatomical connections that permit speech do not appear to be present. Philip Lieberman, a professor of cognitive and linguistic sciences, was the first to suggest that the larynx did not sufficiently descend into the throat of chimpanzees to permit a resonating chamber (Lieberman 1969). Similarly, the human newborn is not capable of speech as much because of its anatomy as anything else. Its hyoid bone—the little u-shaped bone sitting above the larynx—and the larynx itself are high in the throat as they are in other mammals. Over the course of infancy, however, they gradually move down the throat. Through this process, the human vocal tract develops into a double resonating system that is as long as it is wide. This structure, along with the mobility of the tongue (apes cannot lick the back of their molar teeth, for example), is the basis for human speech (Nishimura et al. 2003). While the larynx does descend in chimpanzees early in infancy, the hyoid bone does not. Hence, the ability to make a range of articulate sounds is not present. The apparatus used in making speech is appropriated from swallowing, breathing, or locomotion.

Evolution of the vocal apparatus is thought to have occurred in two phases, resembling the growth pattern in modern human infants (Nishimura et al. 2003). The speech-critical phase of the descent of the larynx was accompanied by changes in the skeleton of the face and lower jaw, which would have increased the resonating chamber as well as the mobility of the tongue. The tradeoff was an increased risk of aspirating food; you can't talk and swallow, or swallow and breathe. These skeletal changes were accompanied by what is called "flexion of the cranial base"; that is, from a skull jutting outward, the bottom of the skull containing the major hole through which the spinal cord emerges from the brain moved anteriorly, toward the chin. At the same time, the face was gradually coming to a position underneath the eyebrows, reducing prognathism. Suggestions as to why this occurred include the reorientation of the body due to bipedal posture and the change in diet that accompanied the increase in body size, which then required a change in the swallowing mechanism as an adaptation to the new diet. Meanwhile, the search for genes involved with, enabling, or controlling human speech is underway.

Recent research has shown that the neuronal precursors of speech are, however, found even in monkeys. Called *mirror neurons*, these are neurons in the brain that are activated when a monkey grasps or manipulates an object. More significantly, these neurons fire when a monkey even sees another monkey. The neurons are thought to facilitate communication since both the sender of a message and the recipient must understand the sender's message (Ramchandran 2005). The mirror neurons are a mechanism allowing for the sharing of meaning and hence are important in the evolution of speech (Rizzolatti and Arbib 1998). The neurons link to an area that is part of Broca's area on the left side of the brain, which, along with Wernicke's area, is traditionally associated with human speech. Recent experimental evidence now indicates that the connections in the left side of the brain, both the traditional language areas and others, are found in chimps as well as humans. Moreover these parts of the brain are active when chimps are communicating, manually or with sounds (Taglialatela et al. 2008). Pointing and gesturing are usually done, as with humans, with the right hand, hence activating the left brain. Under experimental conditions, chimps made new sounds to get the attention of human experimenters working with them and would change the nature of the vocalization depending on the situation. This means that chimps not only are signaling intentionally, but that the referent—the thing referred to in the communication—is understood by others (Taglialatela et al. 2008). These authors conclude that the development of speech undoubtedly evolved from a manual, gestural communication system sometime before chimps and humans diverged. Increasingly, new imaging technology for the brain shows us that language has distinct "architectural" anatomical features. We now know that parts of the brain dedicated to vision in the macaque have become appropriated for understanding the meaning of words, while after the chimp-human divergence, two regions in the cortex became disproportionately large, and new connections were established between old areas and language centers. Language development was not a by-product of brain size, but reflects systematic selection for communication (Rilling et al. 2008).

There are others, of course, who maintain that the evolution of speech depended on the vocal-auditory channel—that is, that speech must have developed from sound systems already in use. The direct antecedent may have been baby talk or "motherese" (Falk 2004). Parents talk to their babies in a singsong way. This musical speech creates a framework on which baby learns to talk. About thirty years ago, scholars noticed that babies move in rhythm with the prosody of speech directed to them. This "musical speech" became a template for the language the baby was going to learn. Linguists have identified the gradual acquisition of new sounds over time. Early human speech would not have resembled ours very much. The question is what subtleties may have been exchanged in the primordial speech system.

The reception of speech depends on hearing. Evidence from fossils in Spain, dating from *Homo erectus*, affirms that the hearing of our Ancestor was not like chimps, who hear in the 1 kHz to 8 kHz range. At 5 kHz, *Homo erectus* was hearing rather like contemporary humans, who hear in the 2 kHz to 4 kHz range (Martinez et al. 2004). Speech and hearing had to have developed in synchrony. A particular gene, *TECTA* (alpha-tectorin), is implicated in this evolution. Its protein product is involved in the functioning of the membrane of the inner ear (Clark et al. 2003).

Genetic research may elucidate whether there are genes for language. The debate centers on the hypothesis first proposed by Noam Chomsky in 1953. Recognizing that children learn language apparently without instruction, Chomsky argued that humans have ingrained a "universal grammar," which is a set of rules about the structure of language (Chomsky 1953). Some scholars consider that there must be multiple genes controlling language development. In contrast, in a recent study, the gene *FOXP2*, which codes for lungs and brains in mice, was isolated as possibly being responsible for language development in humans. It is a member of a group that switches other genes on and off. *FOXP2* was located after a search for the cause of a specific language disorder that ran in one family. It operates in the embryo and seems to pattern the brain to enable it to make language (Vargha-Khadem et al. 2005).

Currently, it seems clear that human language developed to enhance simple communication, but it increasingly became a vehicle to refine thoughts as well (Jackendoff 1999). The primary step was for a sound to have a referent. Monkeys can do this. Years ago, Tom Struhsaker noted that vervet monkeys in Kenya were able to differentiate, by single sound units, between danger in the sky and danger on land (Struhsaker 1967). Monkeys cannot, however, express that the danger has come, will come, or ask if anyone believes it is imminent. They are stuck in a very "now" time frame. Human vocabularies far exceed the repertoires of discrete sounds used by monkeys, and even by apes, and require quick recall as well as long-term memory to store them all.

A most important aspect of human language is its generative ability. From even a very few elemental sounds, humans can generate hundreds of words and even formulate new ones. A study, published in 2008, experimentally demonstrated that forest monkeys *combine* sound units into meaningful statements, adding information in a sequence that details the vocalizer, the thing about which the individual is vocalizing, and the direction of movement in response to the threat (Arnold and Zuberbühler 2008). This is a sophisticated ability long considered to be out of their cognitive realm and clearly implicated in the origin of speech. When the human child is learning a language, she first decomposes sound units and then builds new ones. Equally impressive in a child's speech is the learning of grammatical rules and the misuses of these in pursuit of "grownup" talk. Hence, we might hear things like "I gave the cat food," and the like, or, more interestingly, "the car smushed the squirrel" (from crushed and smashed).

The concatenation of two sounds with referents was undoubtedly an early step toward human speech. "Eagle come" or "snake there" would be powerful additions to a simple ejaculation of "eagle!" or "snake!" Word order would have become important as utterances got longer. "Me kill snake" is completely different from "snake kill me," and hence rules of word order had to be developed. And then, in perhaps accelerating increments, adverbs of place and, above all, time would have developed as terms of spatial relationships and quantifiers (*more, less, to, from, up, now, tomorrow*) (Jackendoff 1999). The point here is that with very little linguistic competence, but with the bare beginnings of protolanguage (Bickerton 1990), hominins would have had greater intellectual range than contemporary apes. Once again, language development demonstrates the rule that *more requires still more*; a new development in human evolution became itself a selective pressure for structural and physiological adaptations.

The information and discussion of current research on nonhuman primates provided in this chapter make it clear that life for our nonhuman cousins is not simple, reflexive, nor "instinctive." While they do not send kin to Mars, sculpt, or write poetry, nonhuman primates do live complex emotional and intellectual lives, ever more complicated by the pressures on their environments and on the daily requirement to find food. From our knowledge of nonhuman primates, we can virtually reconstruct what our Ancestor may have looked like—at least in general form—and similarly, how he behaved. We can assume, for example, that males were larger, taller, heavier, and more robust than females because this is still a trait in some Old World monkeys like baboons and, of course, is common in the apes. The difference in body shape extends to the jaws, canines, heads, and even the crests on the cranium, with the males more robust in all these features. It has long been assumed that this sexual dimorphism, where male and female have different body forms beyond the obvious genital differences, was selected for inasmuch as males defend females and groups, and females defend the young and themselves. Over the course of becoming human, however, dimorphism diminished. While males of our species, *Homo sapiens*, may generally be taller and more robust than females, there is enormous overlap of the normal curves of distribution for height and shape for both sexes, with some females being larger than the average male.

The Ancestor might have slept in the trees or formed nests on the ground and, like contemporary chimps and Old World monkeys, fed from the ground, from under the ground, and from in the trees. They probably lived in wooded areas, but ranged across even more open habitat to forage. The meat and insect component in the diets of chimps and in the diet of monkeys as well, suggests that insect eating played a large part in the evolution of humanity. The Ancestor would have lived in groups, but the nature of these groups is uncertain. Like chimpanzees, there may have been different units within the group depending on function: sleeping groups, mother-infant units, breeding pairs and the like. They lived by their traditions, based on cognitive processes, among which communication, innovation, and curiosity were paramount: processes that I argue are vital in explaining the first approach to and increasing mastery over fire.

No Place Like Home

Evidence of Fire and the
First Use of Hearths

You would think that something as important as *the* major technological change in all of human history would be well discussed in all its facets in books exploring the evolution of humankind. Yet there has not been much attention paid to the importance—especially the physiological importance—of fire in the lives of early humans. The lack of evidence of any hominin relationship with fire prior to 1.6 mya might account for the absence of discussion. No evidence, no discussion. Even when there is absolutely clear evidence of fire, there is still little consideration given to the pervasive importance of this technological achievement. Little wonder, then, that it is generally inconceivable to think that any hominin before *Homo erectus* emerged—at close to 2 mya—could have been capable of capturing and using fire.

What is generally considered solid evidence of fire exists where the remains of home bases with proper hearths have been found, known from a strong concentration of burned pebbles and charcoal—two essential proofs of a purposeful fire. The evidence for these hearths clusters around 400 kya

(Rolland 2004; Gowlett 2006). In the absence of true hearths, however, patches of baked earth, discoloration of stone or bone, rough texture, and distinctive fractures produced by extreme heat on these materials, are indications of fire use, and certainly fire was "captured" long before the human ancestors of 400 kya had any idea how to make it (Stewart 1958; Clark and Harris 1985). It is also unlikely that earliest fire use would resemble the manufactured fires of the Upper Paleolithic, roughly dating to between 40,000 and 10,000 years ago, or even the hearths of the Middle Pleistocene, beginning somewhere around 400 kya. Because the earliest fossil finds come from open sites, material that the hominins of that time used would not be confined in any structure and less would survive to fossilize. Even more contemporary gatherers and hunters were known to make "casual" fires as they moved from area to area; that is, lighting grass or a few twigs, feeding it for awhile or overnight, but not making proper hearths—a specific place bounded by something relatively heat resistant, such as stones, or dug into a pit where a fire is ignited. Had there not been an observer present to document what the people were doing, no trace of these casual fires would have been left. Open-air fires are rarely enclosed. The result is that the cooked remains of plants and animals intermix with ground deposits. As the makers of the first fires searched in them for roasted foods, they displaced cinders, ash, and fire-cracked stones, causing the center of the fire to drift and making it difficult to locate (Asouti and Austin 2004). Moreover, erosion by wind and water makes the likelihood of finding, let alone recognizing, a hominin fire from these early dates improbable (Straus 1989; Klein 1992; Asouti and Austin, 2004).

Many fossil sites are located near water, where perhaps the hominins stayed for a few days, using materials nearby to keep the fire going. Or perhaps they left after only one night for reasons we will never know. Perhaps each site was on a circuit and they returned to the same area after several days. A bonus of this routine would have been that any fecal matter deposited would have dried and been affected by UV radiation, reducing the chance of parasite infestation. Baboons are known to use trees on a 4-day cycle; their waste dropped out of the tree is virtually disinfected before they return again (Hausfater and Mead 1982).

Ash, blackened earth where fire was made, or the charred remains of bone or stone are the types of evidence we might look for to be sure that a fire was used by the Ancestor. Clearly we could not expect a fire*place* at such an early time; even a proper hearth, an area surrounded by stones or dug into the ground at the site, is a little out of reach. These, too, are relatively recent innovations in human history. Hearthlike structures—a dug pit or a circle of stones—are securely dated to the Middle Pleistocene, about 400,000 years ago. Archaeologists and anthropologists have also found circumstantial evidence for the use of fire in the size and shape of hominin teeth, which became smaller and lost their tool-like function and shape, and the association of charred materials with hominin bones, suggesting the possibility of use of fire at early horizons. The burnt bones found recently at Swartkrans, South Africa, dating from 1.5 to 1 mya, affirm such a relationship with fire and are associated with *Homo erectus* (Skinner et al. 2004).

What is wrong, then, with dating the first use of fire to *Homo erectus*? The timing is right for the remains associated with the charred material at Swartkrans, and there is increasing consensus forming that this very humanlike hominin was present between 2 and 3 million years ago, distinguished from earlier forms by its anatomy: teeth, jaws, and muscles. Some scholars suggest—and perhaps not so facetiously—that put into a hat and coat, this person could wander any metropolitan city and not raise an eyebrow. One scholar (Ambrose 2001, 1748) describes this being in the following terms:

> Stone tool technology, robust australopithecines, and the genus *Homo* appeared almost simultaneously at 2.5 MA. Once this adaptive threshold was crossed, technological evolution was accompanied by increased brain size, population size, and geographical range. Aspects of behavior, economy, mental capacities, neurological functions, the origin of grammatical language, and social and symbolic systems have been inferred from the archaeological record of Paleolithic technology.

Of course, there are differences between this person and us, but to the extent that Ambrose's description is accurate, this person is very, very far from a chimpanzee. Not so the Ancestor.

Although it would be acceptable to place the whole discussion of fire's effect on humanity at this more recent date of 2 mya, I am reluctant to do so because the specializations that are apparent with *Homo erectus* indicate a being so nearly us that the process of hominization is being *finalized*—not started. I cannot emphasize this too strongly. Our contemporary notions of progress incline us to expect technological and social change to occur rapidly, and to have done so always. So it is clearly surprising to us that our ancestors were so conservative in their ways that they would retain the same shape of a tool for millions of years, even as biological change was transforming them into a new species. And they did this *twice*: once with the early flake and "cobble" tools of Mode 1 (what used to be called the Oldowan), and again with the bifaces of Mode 2 (the former Acheulean) types. Even anatomically modern humans did not manifest "modern human behavior"—knives, jewelry, or drawings on cave walls—until thousands of years later, that is, toward 50 kya (McDougall et al. 2005). "If it ain't broke don't fix it!" takes on a whole new depth of meaning in this context.

We have noted that the presence of a strong concentration of burned pebbles and charcoal at a site provide two essential proofs of a purposeful fire. But at the earliest horizons, before *Homo erectus*, and in which our early hominin ancestor was just beginning to control fire, finding remains of stone within a fire site is a chancy thing. As was described earlier in this chapter, early humans lived mostly at open sites, along streams, lakes, or at the forest-woodland verge. Cave habitation came later and only become a regular part of hominin behavior at about 2 mya (Bellomo 1994a, 1994b), which is curious since our nonhuman primate relatives use caves: leaf monkeys sleep in caves in southwestern China, and macaques in Sulawesi go into caves looking for water. Skulls of Gibraltar monkeys have also been found in caves, although how they used the caves is not known. Chimpanzees, too, are known to use caves, probably as shelter from high daytime temperatures (Pruetz 2001). Most of the more secure fire sites are from cave locations, the cave protecting the remnants of the fire. Ironically, *entering* the cave in the first place was probably made possible precisely because of fire. Animals living in caves were no match for a fire-wielding biped. Indeed, remains at Koobi Fora, in Kenya, attest to controlled fire near the *mouth* of the cave, and the location of materials suggest that fire was the center of activities—way back at 1.6 mya. The evidence clearly suggests that the fire at Koobi Fora was used for protection or light, but not for cooking (Bellomo 1994a, 1994b). The use of the deeper interior of the cave is only known from more recent times (Perlès 1977; Rolland 2004) and occurs particularly when ritual and art are part of the culture of *Homo sapiens* in the Upper Paleolithic.

So finding *evidence* of fire cannot necessarily tell us precisely about its *functions* for the beings with whom its remains are associated. Anatomically modern humans, the earliest of which are now securely dated to 195 kya (McDougall et al. 2005), like traditional peoples, used stones to construct hearths in order to roast meat, and even to make steam for ceremonial use. An example of traditional use comes from the Pomo and Miwok peoples, Native Americans of California, who heated stones in the fire and, when white hot, transferred them with two sticks to a pot of water. The heat transferred to the water in the pot, another stone was added, and so on, until the water actually boiled. Roasting, one of the simpler uses of fire, requires heat to surround the food. In some cultures a pit is dug into the embers, and the food is wrapped in leaves and covered with more embers. Sometimes, however, the leaves are not part of the process, the ash on the food making grit that wears teeth down. In recent times, Australian Aborigines made bread this way, placing dough made from wild grains directly in the embers without benefit of wrapping it in leaves. The fire, as an instrument for transforming raw plant material that may be inedible into "food," was used as casually by contemporary peoples as it may have been in the distant past. Food animals were tossed without ceremony onto embers, allowed to char, removed, rubbed, and then eaten. There is an interesting account first published in 1921 by George Thomas Basden (2001), who traveled among the Igbo of Nigeria, which underscores how casual "cooking" can be.

> I was busy at work one afternoon when some men began to run
> and shout, and soon afterwards two of them came along with a
> slim black snake, six feet in length, killed in the act of stalking a
> chicken. The two men moved a few yards away, collected some
> dry grass and started a fire. The reptile, balanced upon a stick,
> was thrust into the flames and smoke until it was charred, but
> by no means cooked, and the men promptly devoured it.

Food could be made with little preparation, ceremony, or fuss, or, conversely, with tremendous care. Here is how one archaeologist (Binford 1983, 167) described the way Australian Natives cooked large mammals.

> The burning wood is flamed up to a fast burn. Singeing the game as well as occasionally beating the burning wood results in the accumulation of a substantial bed of charcoal. Once it is judged that enough charcoal has been scaled off the burning wood, the remaining burning sticks are pulled out and tossed to the side, leaving only the charcoal in the pit and on the platform. . . . The kangaroo is nested in the charcoal within the pit, followed by the birds wrapped in leaves to hold in the juices formed during cooking. Once the hot sand and charcoal from the platform are shovelled into the pit to cover the meat, the cooking begins.

The fire itself was so important that it was tended and maintained rather than begun anew. Among the Efe Pygmy of the Ituri Forest of northeastern Zaire (now Congo), the only time when the campsite hearth was not maintained was in the event of camp abandonment, when fires were left to burn out untended and no one would be there to sweep up and discard the ashes (Binford 1983; Fisher and Strickland 1991).

Unquestionably fire assisted in a range of human activities: cooking, heating, clearing areas, chasing away predators, and hunting (Perlès 1977; Bird 1998; Rolland 2004), not to mention lighting. These activities centered around three types of fire: tree-stump fires, grass and brush fires, and campfires, to which should be added firebrands (Bellomo 1994a, 1994b). Grass and brush fires were (and still are) used to clear areas in the ethnological present to beat down insects and other pests, to produce fresh grass for game (or more recently, livestock), or to maneuver herds to where they could readily be killed (Rolland 2004). Campfires and firebrands give light and have been effective in keeping predators at bay. Campfires can be made directly on the ground, in a pit, or surrounded by stones. They can be made with materials at hand or brought from miles around, depending on the area. The choice of what wood to burn in a fire may depend, in the cultural present at least, on cultural knowledge or even values. However, it is parsimonious to assume, and archaeological evidence supports this view, that even in contemporary horizons people gathered whatever wood was most available, especially if it was dry (Asouti and Austin 2004), even though the various properties of wood were known to them as they are even to chimpanzees (Sanz et al. 2004).

Stump burning, observed in the ethnographic present among Australian Aborigines, for example, was often done, and one of these stumps was observed to continue burning for 8 weeks (Clark and Harris 1985). Spears were made by hardening their wooden points in the fire. When this practice began is of course, not known, but the earliest remains of such spears dates from 400 kya, in Germany (Thieme 1997), although even earlier, in the time of Mode 2, wood is known to have been used by the residue left on stone from the scraping and shaping (Dominguez-Rodrigo et al. 2001). The use of sticks as thrusting or throwing weapons is known to chimpanzees as well as many kinds of monkeys who throw branches down from trees, usually aiming at the observer!

Each of these types of fire leaves a print that is exclusive to it. R. V. Bellomo experimented with fire and found that campfires and hearths, when looked at in profile, are basin shaped and leave burned sediments down to 15 cm or so below the surface (1993, 1994a). Burned tree stumps, however, leave oxidized sediment at the surface and down to 2 cm below. They also leave a characteristic hole in the substrate. Grassfires burn relatively coolly and leave no trace of oxidation, as they move swiftly across the ground (Bellomo 1994a, 1994b). These differences enable archaeologists to determine how the burned remains were made.

Archaeological Methods in the Hunt for Fire Sites

Although archaeology is a very precise science, increasingly dependent on technology and teamwork, the serendipity and intuitive skills of finding a site still exist and continue to be part of the excitement. The area to be surveyed is walked by a team, and, with both luck and foresight operating, a fragment, a token of a previous life—bone, shard, stone, or posthole—is spotted and a site set up. Not all sites, however, have surface evidence, so sites are also found by means of advanced technology such as satellite maps or ground-penetrating radar, which sends electric or magnetic waves through the ground. These waves reflect differences in features below the surface, indicating appropriate locations for further investigation and permitting examination of earth at different depths without digging. Geographic information systems (GIS), now as popular in cars as they are in research, link databases with maps. Researchers create a database in order to associate different sets of data, overlaying features associated with habitation to locate sites with traces of charred materials (postholes, for example).

The site is set up precisely: a grid is established and squares excavated slowly, with all material sifted so that the tiniest clues do not escape. Everything is described and recorded for square, location within the square, and depth. Later this computerized information is used to reconstruct three-dimensional images of what happened at that locale. The beauty of this technique is that, years into the future, someone with an idea that has never been examined can test the hypothesis with this carefully maintained database.

Artifacts can be located at a site even if they are latent rather than evident. The latter are those bits and pieces that have retained their original form—a stone knife or a spear point—long after the site has been abandoned by its original inhabitants. The latent evidence, however, requires greater technology, usually computer analysis, to reveal or reconstruct the features or patterns that have become obscured. A major challenge of investigating the past is the question of when bits of bone, charred earth, charcoal, soil baked by fire, burned food remains, burned flints, rings of stones, or holes in the ground indicate hominin occupation (or "contemporary" human, for that matter), and when they are the result of water, wind, animal agency, or natural combustion bringing them together in what only appears to be a site. What materials are characteristic of a butchery site? A kill site? A living site? The field of *taphonomy* developed in the 1980s as a direct result of controversies in interpretation of such remains. Taphonomy is the study of how what's left of an organism becomes a fossil: how it gets to the location where it is found, what happened to it, and what effect carnivores, wind, or water may have had. Flowing water, or even just plain gravity, can move objects that have been altered by heat (an artifact attesting to fire). Sediment covers the artifacts, and tree roots or burrowing animals can move them or even distort their shape (Petraglia 2002).

The field of taphonomy was launched by experimental studies that tested the processes thought to be involved in fossilization. Taphonomists experiment with materials and conditions to try to understand and reproduce the events that formed the locale in question. These researchers have designed ways of telling whether fracture patterns on stone were made by stones bumping against each other or by human hands. By using the electron microscope, they can tell if cut marks on bone came from stone tools or carnivore teeth, and by using statistics and demography, they can tell whether a given bone accumulation or distribution occurred due to human, animal, or natural agencies. Predators, for example, tend to target the young, old, and infirm—those easiest to isolate from the herd and take down. Human hunters, however, tend to take animals that provide the most food for the effort. Group hunting and strategies such as using inukshuks—stones piled up to look like people—help in procuring large returns. Accumulation of young animal bones versus accumulation of old or mixed bones, therefore, can indicate who lived in that spot and for how long. Past environments are reconstructed—temperature, flora, and fauna are assessed and the site put into a meaningful context. Tiny things tell a story: pollen can be identified so that the ecology of the region can be reconstructed. Knowing what plants, trees, and shrubs were in an area indicates what the fossil beings at that locale probably ate or used for medicine (Brain 1993b). And we know they must have used plant material for medicine since we know a variety of vertebrates use plants—or even soil—for medicine (Bolton et al. 1998; Burton et al. 1999; Engel 2002; Klein et al. 2008).

With the newest molecular techniques, the diet of the animal whose bones were found at a site can be reconstructed on the basis of the ratio of two carbon isotopes: ^{14}C to ^{13}C. Bits of bone can be analyzed to compare their protein sequence with known standards to determine what type of animal this bone belonged to. This study can be extended to periods much earlier than DNA analysis can interpret; hence, it is critical to the study of early sites and the evolution of fauna, including humans.

Even when debris is *known* to have been made by controlled fire, it may be nearly impossible to demonstrate that fact. Just as chimpanzees leave sticks at a site, which are subsequently used by other visitors (Sanz et al. 2004), so too did hominins reuse materials that were left by earlier occupants at a given site. Overlaying one use over another makes determining who used it and for what purpose a most difficult job. The conservation of materials this multiuse suggests is reflective of the availability of materials, the appropriateness of materials, and perhaps even the general primate inclination not to overdo when it is not really necessary! (Monkeys would rather raid a plantation for fruit growing in neat rows, despite the risk and experience of getting injured, than work foraging in the forest gathering fruit from several trees.)

Archaeology is now a team effort incorporating the knowledge and skills of earth scientists, geologists, foresters, and botanists, among others, who use sophisticated chemical, biological, and geological techniques to understand the history of an area. The chemist or physicist can reconstruct the temperature at which the artifacts were heated. However, while *thermoluminescence* can date burned materials, it can only go back as far as a 250 kya. It depends on the accumulation of radioactivity, specifically electrons trapped in crystals. Artifacts give off a flash of light that is proportionate to the amount of radiation that was trapped in the material, as well as dependent on the length of time since that material was heated. The lapsed time since the radiation was deposited is figured out by measuring the total amount of stored thermoluminescent energy (the paleodose) and the annual radiation dose (ARD). This is the rate at which the thermoluminescent energy is accumulated, and a calculation of the time since deposition can be made (Price 2003). To date objects that are even older, scholars have turned to *electron spin resonance*, which measures the spin in a magnetic field of free radicals (Skinner et al. 2004). Free radicals are atoms that are not joined to other atoms at a given moment and so are able to take part in another chemical reaction. They are produced by different processes, such as radiation, damage, or fire reaching a temperature of about 600°C, as is found in a campfire. Although the technique is mostly used on nonbiological samples, it has been used to date bone, as free radicals become embedded in bone as it becomes fossilized. The number of trapped electrons can be counted in bone, teeth, and shells, and dating is good to 1 mya. As these unpaired electrons are highly reactive in the biological context, the sample must meet exacting criteria before it can be used for dating (Grun et al. 2005).

Table 6. South African sites of hominin remains		
Location	Hominin	When (mya)*
Sterkfontein	*A. africanus; Homo erectus*	From 3
Makapansgat	*A. africanus*	3–2
Taung	*A. africanus*	3–2
Swartkrans	*A. robustus, Homo erectus*	2–1
Kromdraai	*A. robustus*	2–1

Note: * Unit is in millions of years ago.

South African Archaeological Sites

Of the South African archaeological sites important to our search for the earliest uses of fire, there are three major ones among thirteen: Sterkfontein, Makapansgat, and Swartkrans. Two other sites have also yielded important information: Kromdraai and Taung. Table 6 summarizes the location, the approximate age of the remains, and who was found there.

The oldest location is Sterkfontein, biostratigraphically dated at 3 mya. Makapansgat and Taung are next, dating to between 3 and 2 mya, while the final two, Kromdraai and Swartkrans, date to between 2 and 1 mya. Dating of the South African sites is difficult because of the complex geological history. Paleomagnetism and biostratigraphy have proven useful. *Paleomagnetism* works on the basis that the magnetic poles of the earth have shifted 60 times over the past 20 million or so years (Gubbins et al. 2006). Now the poles are northerly (normal), so the former southerly orientation is considered "reversed." This normal condition has been around for about 780,000 years and the reversal preceding that one occurred around 2.6 million years ago. Northerly orientation had another episode around 3.4 million years ago, and the beginning of the reversal before that is not yet known. As igneous rocks are formed, the magnetic field is "fixed." The record of this magnetism is worldwide, so that rocks from different areas can be correlated in time one to another, and the fossils within the stratum will be from that period (Gubbins et al. 2006).

Biostratigraphy, or faunal correlation, on the other hand, is a comparison of fossil animals of known age with fossil animals of unknown age. Where two strata at different sites contain the same kinds of fossils, it can be assumed that the known date can be applied to the unknown. If an elephant of known age was found at one site with a pig nearby, and a thousand kilometers away the same kind of elephant was found with a horse, it would be assumed that the horse and the pig were coeval—that is, lived at the same time—and that the date applied to the elephant would be correct for both horse and pig. As for hominins, the gracile australopiths (the common name for members of the genus *Australopithecus*), *Australopithecus africanus*, have been found at Taung, Makapansgat, and especially Sterkfontein. The robust form of the genus, *Australopithecus robustus* (also known as *Paranthropus robustus*), has been found at Kromdraai and Swartkrans. *Homo*—although it is uncertain which species (*erectus* or *ergaster*)—has been found at both Swartkrans and Sterkfontein. *Homo ergaster* has been proposed as the name for the African members of the genus—leaving only the Asian members to be called *Homo erectus*—however, they are still being called *Homo erectus* nearly everywhere on the basis that there is little distinction between the Asian and African forms and because these fossil hominins are members of the far-ranging paleospecies whose African form is considered to have given rise to *Homo sapiens* (Asfaw et al. 2002).

Significantly, the robust forms of australopiths came *after* the gracile ones; they are more derived, more specialized to a certain pattern of life, and apparently went extinct without issue, whereas the gracile forms are still thought by some to have given rise to the genus *Homo* (although there are now other candidates). Finding robust australopiths contemporaneous to *Homo* was the clincher; one form is not ancestral to another if they live at the same time or overlap closely in time.

Unlike in South Africa, the East African succession of life forms is dated by potassium-argon (K-Ar). Volcanic rock, found in abundance in East Africa, contains sufficient amounts of potassium to allow this type of dating. This "clock" works on the radioactive disintegration of one substance to another more stable state (potassium to argon, for example). If the right kinds of rock containing sufficient potassium are found at a site, the amount of original material can be estimated and the time it took for it to change to the stable state can be worked out. More recently, different isotopes of argon (^{40}Ar to ^{39}Ar) have been used, especially in South Africa, as they are more accurate, especially where there have been questions of contamination from one stratum to another. Since these radioactive methods do not apply at Swartkrans, as the appropriate materials are not present, the two comparative methods have been used and give a date of between 2 and 1 million years ago, but these still remain controversial.

In 1924 Raymond Dart analyzed a skull that had been retrieved from one of the caves at Taung. At the time of this discovery, Dart was only 32 years old, but he had the courage to suggest that this was indeed an ancestor and named it *Australopithecus africanus*, the southern ape from Africa. *Australopithecus africanus* has been in and out of the ancestral line for decades. Dart had a devil of a time getting the scientific world to admit that the Taung fossil was humanoid, let alone to consider it for a place on the family tree. A critical factor was its brain. The fossil of the brain—the endocast—was thought to exhibit shared anatomical features with humans. Endocasts have been invaluable in reconstructing the evolution of humanity and for assessing capabilities of a fossil based on comparison with the known function of a given brain structure. Now the Taung finding has been reversed. In the endocast of the Taung child, the essential feature is a fold of brain known as the *lunate sulcus*. Dean Falk, a prestigious scholar in paleoanatomy, risked her career to show that this fold had originally been misidentified. This means the Taung child should not be included in the human family (Falk 1992). But by the time Falk determined this in the 1990s, the skull had come to occupy an almost sacred position in paleoanthropology; and disagreeing with "known" information was just not easy to accept.

The intense scrutiny given to this particular fossil is due to when it was found and the emotional investment in coming to consider it related to our human ancestors. Recent studies show that the skull belonged to a 3- to 4-year-old child, and not a 6-year-old, as was first thought. Since its discovery, new knowledge about growth patterns in humans, apes, and monkeys has made it possible to determine the Taung child's age accurately (Zihlman 2004). The correction to a younger age has implications for family life and development. This youngster had apparently been killed in an attack by an eagle (Berger and Clarke 1995; McGraw et al. 2006). I have seen juvenile monkeys in the country parks of TaiPo who have been partially scalped by the claws probably of the resident black-eared kites. Reconstructing what must have happened, it seems the defenders in the group retrieved this young animal from the predators, but the attacks left a permanent mark where the eagle dug its talons into the youngster's head fur, ripping it from its skin so it healed into a flap or ridge. In addition, Dart had evaluated the assemblage of debris of bone, stone, and horn at another site, called Makapansgat, in what had become a limeworks quarry. Dart suggested that these were part of a tool kit used by these early people in their daily lives, and he named this complex the *osteodontokeratic culture*, for bone, tooth, and horn (Dart 1925; Dart 1957). For a good number of decades this "culture" was accepted by some, but the pioneering studies in taphonomy, by C. K. Brain, at the beginning of the 1980s, disproved this thesis. Brain examined a series of South African caves, studied local peoples and how they used resources, especially plants and animal bones, and also observed wild carnivores and primates. His main concern was to investigate how an australopith skull with curious puncture wounds got into a cave at Swartkrans and who made those wounds. Brain matched the holes perfectly to a set of carnivore teeth. He then had the evidence to conclude that the bones of the australopiths had accumulated in the cave not as inhabitants, but as prey (Brain 1981). They had been dropped from a tree overhanging the cave entrance by the big cat that had dragged its hominin prey to that secure height, as do big cats today.

With this knowledge that the australopiths were in the caves as prey refuse rather than residents, the idea of the osteodontokeratic culture crumbled and the idea that these early ancestors could use tools became speculative until decades later when Mary and Louis Leakey discovered Mode 1 tools. (Incidentally, the idea that the early, tiny hominins—none taller than *almost* 5 feet until the advent of *Homo erectus*—were not the predators but rather consistently the prey, has just been developed into a book, *Man the Hunted*.) It makes an enormous difference to view these early ancestors as running from big predators, collaborating in cooperative efforts to keep the predators from attacking them, and inventing materials to help in self-protection against wild cats, hyenas, snakes, and whatever (Brain 1981; Clark and Harris 1985; Klein 1992). Perhaps they even picked up fire sticks to brandish against them. Clark and Harris (1985, 17) ask:

How far back in time are we justified in inferring that hominids were fire users? With this is bound up the question as to how a slender, relatively defenseless, bipedal ancestral australopith was able to be a successful savanna ground dweller and not fall victim to regular predation from large carnivores. Some efficient defense system must have been in operation.

This is no less true of a "slender biped" at 8, 6, or 5 million years ago!

The kit postulated by Dart was really an accumulation of waste due to the actions of various predators and natural factors over the eons. These South African sites—where over millennia caves had closed up, reopened, refilled, and closed up again, to create a most difficult geological situation—have yielded evidence of human culture. Applying sophisticated chemical and histological analysis to bone material that appeared burned at another site at Swartkrans, Brain demonstrated that the bones had to have been heated to high temperatures, which could not happen unless someone had controlled the fire. Experimental evidence came from burning hartebeest bones, recording the temperatures achieved in the campfire, and comparing the results with the fossil material. This method convinced scholars that fire was used in this area by hominins (Brain et al. 1988).

However, the material and its dating remain controversial. The evidence has been dated as being 200,000 to 700,000 years of age (James 1989), but Brain himself places the burned bone at about 1 mya, and, based on the association with hominin artifacts such as tools, he believes that the makers of the fire that burned the bones were *Homo erectus*. But the beauty of science is that it tests and retests. In 2001, scholars tested the apparently burned bone with electron spin resonance (ESR) and concluded that the cave at Swartkrans should be dated to kya, not mya (Curnoe et al. 2001). Even more recently, however, Skinner and her colleagues reexamined the Swartkrans material, also using ESR, but with quite different results (Curnoe et al. 2001; Skinner et al. 2004). They found that the temperatures at which the bones had been burned were too high for grassfire; they had to have come from a campfire, and that means they were burned on purpose. The new date is between 1 and 1.5 mya. Of the two kinds of fossil hominins found in the Swartkrans cave, *Australopithecus robustus* and *Homo erectus*, the fire is considered to be associated with the latter (Skinner et al. 2004). Interviewed about their find for *Science in Africa* (2004), one of the authors said:

> We do not claim that distant human relatives such as *Homo ergaster* or *Homo erectus* were making fires, but at least they were probably using fire in a controlled manner. They could have collected burning branches of trees that had been set alight by lightning after dry winters on the highveld.

The evidence for controlled fire gets earlier and earlier.

A. robustus was a vegetarian, as shown by the large teeth and the markings on the skull showing insertion of powerful masticatory muscles. It turns out they suffered from dental caries—not found in the *Homo* line until the domestication of plants. The robust australopiths seem to have been the gorillas of their time, relying on vegetation, and therefore not as active as the more lithe members of their group, although the hand suggests that they were capable of toolmaking (Susman 1987; 1991).

Brain also found bone that had clearly been used as a tool, dating to around 2 mya (Brain 1993a, 1993b), which is earlier than the hominin-burned remains—if that is indeed what they are. The original interpretation of the bone tool was that it had been used to dig up tubers and that the digger, with its huge grinding teeth, was the vegetarian specialist *A. robustus*. Even vegetarians require protein, of course, and these robust beings—like apes—seem to have gotten a good portion of their daily requirements from insects. How do we know? Further studies of this bone tool show that it was used for gathering termites from their nests, much as chimpanzees use sticks (Backwell and d'Errico 2001). Termite eating among South African australopiths composed up to 45 percent of their diet and was a "fundamental hominin trait," according to Sponheimer and his colleagues (Sponheimer 2005). Importantly, this is *bone* used as a tool, harkening back to Dart's idea of the opportunistic tool kit of early hominins being made from materials at hand. Equally as important, if not more so, is the evidence these authors present that it was termites—not tubers—that were being dug up (Backwell and d'Errico 2001). Tubers contain toxins, some of which are detoxified in the primate body; some, however, remain active even after cooking (Prathibha et al. 1995). Indeed detoxifying this food source is considered by some to be the very motive for cooking (Wrangham 2001). This bone tool, however, had traces on it that clearly showed it was used to dig termites. This means that the australopiths did not have to know what modern humans know about finding hidden tubers or attempting to detoxify them. The use of the tool therefore changes the reconstruction of australopith behavior, as well as the question of the origin of cooking. These ancestors were as capable as contemporary chimpanzees of opportunistically using whatever was available to satisfy a particular need. I found the Sponheimer study incredibly exciting because it meant that these ancestors were using materials at hand opportunistically, and that they were using termites for protein. If apes and recent hominins exhibit the same behavior, it stands to reason that these ancestors would also have been capable of this behavior.

These people, as early as 2.6 mya, were making simple stone tools. The earliest were found at Gona, Ethiopia. The tools are called Mode 1 and are very roughly fashioned. Furthermore, people were making tools some distance from where they were to be used—as much as 10 km away. Not only did early people go some distance to find the material they needed, but they also had preferences for which kind of stone was best suited to their purpose. Originally the large Mode 1 nodules were thought to be the chopping tools themselves, but more recently, as the art of knapping stone has become experimental, it has become quite clear that the nodules were actually "blanks" from which flakes were struck, producing small, sharp-edged "knives." Archaeologists, utilizing the electron microscope, can tell by scratches, dents, and cuts on bone, stone, and even tooth the use to which the material was put. Such marks on bone indicate that tools were used, even if no tools are to be found. Quite simply, nothing else can leave these imprints on bone. At Bouri, in Ethiopia, at a lakeside site, there is ample evidence of marks on fossil bones from 2.5 mya, which indicates cutting, but there are no tools to be found. Microwear patterns on the bones, however, suggest that tools were used to fillet, deflesh, and disarticulate animals (de Heinzelin et al. 1999; Asfaw et al. 2002), and is sufficiently sensitive to show within-species diet variability (Scott et al. 2005). It seems that for our early nomadic ancestors, valuable objects were to be carried around, not left behind—a bonus afforded by bipedality. Of note for the reconstruction of fire acquisition are these facts: (1) hominins carried important things with them; (2) the lack of an object is not a lack of evidence; and (3) nomads do not build lasting structures.

By the next tool technology, one that is associated with *Homo erectus*, the search for appropriate raw material had increased the distance between the source of the stone and the place where it was made into tools to about 20 km (Ambrose 2001). These Mode 2 tools were named Acheulean (or Acheulian) after the site in France where they were first discovered. Over time, and with the advent of *Homo sapiens sapiens*, the distances traveled to find raw materials have been calculated to have reached upward of 300 km (Ambrose 2001). But these were people in a virtually modern context. They were living in the Upper Paleolithic around 40 kya, when *Homo sapiens sapiens* was already well established and using fire to heat structures, to roast meat in proper ovens, to cook in stone-lined hearths, and to create stone lamps (Perlès 1977). Interestingly, the toolmakers at 2.5 mya were also highly skilled. They did not come upon this ability all at once, however. They had to have understood fracture planes, types of stone and how they break, where to place the blow on the nodule, and that nodules do concuss into flakes. If the pebbles to be flaked were too small to hold, they used anvils and broke them with a "hammerstone." Style was not a consideration; these empiricists were using material that would achieve their goal rather than choosing material on which a personal imprint would be placed. When bone was made into tools, it was usually flaked as was stone—if modified at all (Ambrose 2001).

Archaeologists have become quite sophisticated in tracing the origin and development of tool use and have extended their observations and experiments to nonhuman primates. Chimps cannot make stone tools to a regular pattern under experimental conditions. And what they can and do make, as might be expected, is not as sophisticated as artifacts found at 2.5 mya. In an exciting study, archaeologists have begun establishing an archaeology of ape manufacture from the detritus left and partially buried where chimps have been seen to break open nuts (Mercader et al. 2002; Vogel 2002). Nuts are hard to crack and require tools to get at the meat inside. With canines—especially in female chimps—too small to effectively open hard-shelled nuts, tools must be used or the rich nutrient abandoned. The archaeologists excavated the known sites utilizing precise methodology. What they found was bits of rock *used* but not fashioned. Critics have noted that chimps do not choose the stone with which they work; they are opportunists utilizing whatever material they find, yet, as discussed in chapter 5, they *do* choose wood for particular purposes.

For that matter, chimps should not be able to make anything useful at all! The shortness of the ape thumb makes using that digit against the tip of the second digit virtually impossible (Burton 1972a). The long, thin fingers of apes, so well suited to hanging from branches, are an evolutionary distance from the square, blunt fingers of monkeys—the prototype of the human hand. The specializations in the wrists of chimpanzees (and gorillas, for that matter) that are the hallmark of "apes" and that permit the knuckle-walking adaptation preclude the bending of the wrist to bear on the stone; they use their arms from the shoulder since the wrist lacks flexibility (Ambrose 2001). But they *do* make tools, once again affirming that lack of structure does not dictate lack of use, and that function is not limited by structure. In fact, chimps make sets of tools for a specific purpose. Moreover, a recent study indicates that their use of the hand indicates the same kind of mental processes that we have. When we go to grasp something, we take it in a position that will make sense to its end purpose, rather than dictated by the shape of the object or the hand itself—even if that means a somewhat uncomfortable position. This is the "end-state comfort effect," which indicates that a prediction of how the object is to be used is shows how it is meant to be handled (Weiss et al. 2007).

The ancestors at 2.6 mya, whose brains barely exceeded chimp size, fashioned stone tools. And what is even more curious, tool manufacture was still Mode 1 until about 1.5 mya, after the big-brained *Homo erectus* came around. Braininess in itself is not enough, apparently, to cause changes in the way of doing things, although there is increasingly evidence of diversity from site to site. New findings at Lokalalei, Kenya, dating from 2.34 mya, for example, indicate that the local stone knappers seem to have been planning where to get raw materials and how to manage them. In addition, archaeologists consider that the hominins at this locale followed technical rules in a consistent fashion, which resulted in high productivity (Delagnes and Roche 2005).

Four important lessons can be gleaned from these artifacts at Lokalalei: first of all, the ancestors kept important things with them, even though as nomads this would have been awkward at times, not having the proper carrying devices. Second, the same small brain size that enabled chimps to make tools was also found in the earliest ancestor. Third, local exigencies led to the development of local patterns. Fourth, retention of a way of doing things has gone on for millions of years in some areas with little change.

However old the human *use* of fire may be, it certainly accelerated before 1 mya—at least as far as can be measured from carbon remains of a particular sort from sub-Saharan Africa. There are some very early dates for materials deliberately burned by fire—back to 2 mya. However, scientists are careful to differentiate what *may* be the use of fire from what is *clearly* fire use by hominins. Table 7 marks with an asterisk those sites that are currently considered true sites of burned artifacts with secure dates around the 1-mya mark.

Not included in table 7, because it is too uncertain, is a site from the Middle Awash of Ethiopia, dated to between 2 mya and 500 kya. The evidence of fire use at this site consists of cone-shaped, reddened patches of clay, 40–80 cm in diameter, associated with stone tools, and cranial fragments of *Homo erectus*. The clay was found to have been baked to 200°C or perhaps more, but could not have been the result of lightning strikes producing what are called "fulgurites"—baked clay (Clark and Harris 1985). However, the other archaeological materials found near the site were not directly associated with the clay patches. Judging from other sites, like a hippopotamus butchering site, stone tools and clumps of burnt clay ought to have been in close proximity. On the other hand, it may well be that the burnt clay was the remains of burnt tree stumps that had termite mounds at their base. Hominins as capable as *Homo erectus* may have set the tree stumps ablaze to have a source of fire (Clark and Harris 1985). It is clear that the fire would not have been used to cook meat, as the faunal remains nearby were not burnt.

Controlled use of fire is clearly demonstrated at Koobi Fora, in Kenya. The hominin responsible is variably thought to have been either *Homo erectus/Homo ergaster* or *Homo sapiens*. The most difficult procedure for archaeologists working at early time horizons is to find methods that

Table 7. Sites of burned artifacts with dates			
Area	Site	Country	Date
North Africa	Koobi Fora	Kenya	*1.6 mya
South Africa	Swartkrans	S. Africa	1.8? mya *1–1.5 mya
East Africa	Chesowanja	Kenya	*1.4 mya
Middle East	Gesher Benot Ya'aqov	Israel	*790 kya
Shanxi Province	Xihoudu	China	1.8–1.0 mya
Yunnan Province	Yuanmou	China	1.7–0.6 mya or 600–500 kya

Note: Date units given in millions of years (mya) or thousands of years (kya).
Note: *Asterisk indicates those sites considered true sites of burned artifacts with secure dates around the 1 mya mark.

allow them to evaluate whether or not what appears to be an artifact made by someone actually is one, and, if so, whether the fossil associated with it is the manufacturer. At Koobi Fora, at a site dated to 1.5 mya, stone tools and the jawbone of the robust type of *Australopithecus* were found. Was this robust form the maker? More likely it was made by an early species of *Homo* (Phillipson 2005). At an earlier site, a different sort of problem emerged: dated to 1.6 mya, bone was found whose color suggested it *might* have been burned. If that were the case, it would be evidence of cooking. But the original analysis suggested that the bones had turned whitish due to weathering, and not human agency. The earlier site also yielded several similar, apparently thermally altered artifacts. The effect of heat is to break, crack, or discolor objects, so it was assumed that the stone material found was not significant. More recently, however, this discolored stone material is thought to have been the result of humanmade hearths. Sophisticated statistical techniques were employed to separate the stones into significant and insignificant groups. These were subjected to rigorous analysis, each bit catalogued, placed in a grid, the distance from other bits calculated, the cluster of bits considered, the density established, and, finally, the probability that these pieces were randomly distributed rather than purposefully placed was measured. The results suggested that, yes, there were tools amidst the debris (Bellomo 1994a, 1994b). And not only tools! There were actual hearths that had been repeatedly used and heated to between 200°C and 400°C. This was determined from discolored patches of sediment, reddish orange in color, 40 cm in diameter, and up to 15 cm in thickness (Bellomo 1994a, 1994b).

Recent experimental studies in archaeology demonstrate that beneath the campfire, temperatures can range from 100°C to as high as 700°C (Wertsa and Jahren 2007). The techniques employed enable archaeologists to determine whether traces of fire were deliberately managed or accidental (Linford 2001). At Koobi Fora, the evidence indicates that fire was used and controlled. However, its function was not what you would expect. These people were not using fire at their hearths to cook. Rather, the evidence, at least at Koobi Fora, makes it clear that, even as recently as 1.6 mya, fire was being used as a *light* to see by and perhaps as a means to keep predators away. In terms of my thinking about the Ancestor, this is an important supposition because it demonstrates that fire has differently perceived functions. Because it gave off heat did not necessarily mean it was used for cooking. Hominins would have husbanded this resource for the simple reason that predators would not enter a lit area and because the fire itself, not to mention its heat and its awe, could be used as a weapon and to keep them away. These early uses would mean that hominins did associate with fire and would have been exposed to its effects.

The South African site introduced earlier, the cave at Swartkrans, supports the contention that early hominins, somewhere between 1 to 1.5 mya, were systematically using fire (Brain and Sillen 1988). There were nearly 60,000 fossil fragments found in the cave, including 270 burned animal bones and numerous Mode 2 tools. Importantly, there were animal bones with cut marks indicating butchery, some of which had been fashioned into bone tools. In the late 1980s these materials were thought to be much more recent in time, but studies with newer technology have affirmed the earlier date (Brain 1993a, 1993b). There is still some doubt as to whether or not the bones were burned by human agency or whether some other explanation will come to light. At the time of their discovery, the authors experimented with different temperatures of campfires and different vertebrate bones in order to see which of the experimental results most resembled the discovered bones. The higher the temperature, the more likely the fire is to have been nurtured by human agency. Results suggested that 46 of the bones were lightly heated, that is to temperatures below 300°C; 52 bones were heated to between 300°C and 400°C; 45 bones to 500°C; and 127 bones to beyond even this high temperature. The association between burned material and human occupation is, for many scholars, sufficiently strong to confirm that fire was being used (Bird 1995). Note that no one is yet suggesting that fire was being *made* at this time horizon, just used. Granting that natural burns reach high temperatures as well, it is important to note that the context is different. It takes large spaces and a great deal of fuel to bring natural fires to these temperatures, whereas the evidence indicates that humans were nurturing fire in a confined area and under their control.

Thus, the earlier the date, the greater the ambiguity. At Chesowanja, in Kenya, at a horizon securely dated by potassium-argon (K-Ar) to 1.4 mya, stone artifacts and a series of baked-clay clasts have been found. Are these remains a function of a fire made by early people? It seems likely, as there is also what appears to be an actual hearth: an arrangement of stones that had been placed around the fire. There were fifty-one clay specimens, all baked to a reddish-brown color (Gowlett et al. 1982). They ranged from little flecks to substantial lumps of about 5–7 cm in diameter. In general, they are rather like similar materials found at Koobi Fora and other sites in East Africa. Subjected to archaeomagnetic analysis, their antiquity was confirmed—they were from the same time as their surroundings. However, their manufacture remains in dispute. Some scholars have felt that bush fires could have caused this earth to become baked, but the original researchers are positive that these clasts are a function of human-controlled fire—in particular, a slow burning fire that achieved temperatures of at least 400°C–600°C, the temperature experimentally known for open campfires. Furthermore, the baking of these lumps would have taken longer than a wildfire would have allowed. Needless to say, there has been a great deal of controversy over the structure and the burned material. If the material at Chesowanja does represent a hearth, it will mark a very early time indeed for such human activity.

Control of Fire and the Development of a Home Base

A hearth, as a central place around which a family sits, sleeps, and interacts, seems a logical place to begin considering a "home base." Monkeys do not have a home base as such; neither do chimps. Home bases are a cognitively sophisticated concept—what some have called an *emergent* behavior—that is, one that has no *direct* primate antecedents (Ronen 1998). Home bases are characterized by having a fixed location and by being the place where foods are shared and where the group sleeps and transmits information through its socializing (Rolland 2004). But monkeys and apes do have three-dimensional home ranges and, in those spaces, we can identify *core areas* that are the zones in which they spend most of their time. These areas are well known to them, and that knowledge overlaps with the functions of the home base. The core area is considerably larger than a home base—and that makes a substantial difference. They return night after night to these sleeping areas, season after season to these fruiting plants, trip after trip through the same branches or along the same well-worn paths (as do other mammals). Important features like predators, food, and water are predictable on the basis of past experience. Monkeys will stay in the same fruiting trees for several days, dining on ripe fruit, and then move on when they have exhausted that supply. Gorillas and chimps usually make new nests each day (Banjjo et al. 1981; Groves and Sabater Pi 1985), even if they return to the same site. They can and do, however, occasionally reuse the same nest. In a real sense, then, while there was no "home" as we understand it, the whole range that the Ancestor visited served that function.

To the more terrestrial *Homo sapiens*, the space above our heads has no real value; to an ape or monkey, on the other hand, it is an integral part of their thinking. Their mental map of home, in this sense, would include every possible snake and predator, every potentially fruiting bough, every sleeping branch—a three-dimensional world. Assuredly, the Ancestor knew the terrain that his group traversed as well, if not better, than does a monkey or ape. But a home base is more: size is one factor, and, more importantly, there is a great deal of *choice* involved in where to locate the home base. It stands to reason that the habit of gathering near a fire would promote what becomes the home base. If hominins were seeking out burning stumps or other forms of natural fire in the millions of years before the Middle Pleistocene, and subsequently carrying fire to ignite their own at whatever place they chose to settle for a night or a day, the tradition is established. To my thinking, the development of a home base is a natural outgrowth of the preexisting relationship with fire, and evidence suggests that the variables being computed reach a critical mass around the Middle Pleistocene or a fraction earlier. The most impressive and securely dated site for a home base is the one in Israel, lying in time between the Middle Pleistocene and 1 mya. In this very early horizon, people were members of our genus, *Homo*, living the fundamental pattern of hunting, gathering, and socializing in and around a fixed location, the home base. The origin of the home base stems perhaps from the primates' core area, but develops with the increasing control of fire. This progression moves from fire as the center of a camping site, occupied for a brief period, to pits or a place surrounded with stones. These are proper hearths, and at this point there is now a true home base. The progression continues, and eventually the place in which the hearth is located (usually caves) has become *home*. Ultimately the very notion of home as concept and structure becomes independent of fire.

The social and cognitive implications of a home base indicate a being very different from its forebears. A good number of definite hearths have been recognized beginning after 500 kya, mostly around 400 to 350 kya. There are other sites that remain controversial even at these recent dates. The hearthlike arrangement at Chesowanja at 1.4 mya, therefore, certainly incited discussion. According to some scholars, that structure may have been formed by water flowing in a small channel caused by high rainfall (Isaac 1982). However, sorting of stone artifacts should also have proceeded according to size along the furrow, but this was not the case. The faunal remains, independent of their size, were found randomly throughout the site, and hence were not accumulated by fluvial action either, as water transport would be more likely to concentrate objects of similar size or weight. Furthermore, the clay lumps, which have retained sharp protuberances, could not have been transported by water as the edges would have been dulled (Gowlett et al. 1982). In addition to these artifacts, natural stones were found that appeared to be in a hearthlike arrangement.

Of course the closer in time, the more secure the interpretation. Recent Israeli finds come from the Hula Valley and have been dated to

790 kya. The material from this site securely indicates the use of fire and is the first absolutely clear-cut demonstration of such behavior by hominins. The inhabitants of the site were collecting plant food and processing meat that had been hunted. Equally important are thousands of specimens, including fragments of flint, fruit, grains, seeds, and wood, whose distribution was carefully mapped. These burned remains are all identifiable, three of which might have been part of the diet: olive, wild grape, and barley. All the material was divided into two categories: burned and unburned. Although only 2 percent of the flint and 4 percent of the wood had been burned, the distribution of these at the site was significant. While the burnt relics are gathered in special locations, they are not the hearths of more recent times, although this site comes very close: charred flint fragments suggest that burning was done in specific places. Had burning been the result of natural causes, there would have been more charred remains, and there would have been a more even distribution of burned fragments throughout. However, the localization of burned materials suggested that the fires were intentional, the material to be burned was chosen, and the fire controlled (Goren-Inbar et al. 2004). The authors of the burning, however, are not yet known—they might have been *Homo erectus* or even archaic *Homo sapiens* (Goren-Inbar et al. 2004).

Homo erectus was a cosmopolitan hominin. Remains are found in Africa, Europe, Eurasia, and Asia, and it is thought that northward movement out of Africa was made possible because fire was part of the inventory of their culture. Northward expansion into cold areas (as well as into the depths of caves) became increasingly possible as this species learned to control its own environment. The earliest remains of fire in Asia come, at the present, from China. (I say "at the present" because there is a bias in archaeology as a function of *where* exploration can and does take place. Political and financial issues are pertinent here. The Cultural Revolution of the 1960s seriously impeded archaeological—or any—research in China, for example.) The Chinese material shows the care and sophistication now possible in assessing ancient discoveries. The earliest and currently most secure findings in China come from Zhoukoudian (formerly Choukoutien) (Rolland 2004).

Medical doctors, very much like Sir Arthur Conan Doyle's Dr. Watson, being detectives in any case where human illness is concerned, are often intrigued by early human history and have been important allies in the discovery and analysis of fossils. This was certainly the case for Davidson Black, a Canadian medical doctor, comparative anatomist, and dedicated student of human evolution. When the opportunity arose to work in the Peking Union Medical College in China, he happily took the post in order to be able to search for Ancestors in what was then considered to be the "cradle of humankind." In 1929, a geologist, Wenshong Pei, was involved in an excavation at Zhoukoudian. When his team found a skullcap 30 meters down, the excitement apparently was almost more than anyone could bear. He brought the skullcap to Davidson Black for analysis. It took Black almost four months to extract this bit of skull from the matrix in which it had so long been buried. Thus was born *Sinanthropus pekinensis*, later brought into the fold of *Homo erectus*. The saga continues, and it has been beautifully narrated by Harry Shapiro in *Peking Man* (1975), telling how at the beginning of the Japanese invasion of China in the 1940s, the bones of this important fossil were to be sent for safekeeping to New York, but never got there. They were somehow lost at the dock before loading and have never been recovered. Fortunately, casts had been made of the fossil and it is these that originally formed the basis of our knowledge. A committee was formed in 2005 by the Chinese government to locate the fossils.

At the time of the find, Black began investigation of additional bones from the site that appeared burned, were distorted in shape, and discolored to black, gray, blue, and bluish green. Chemically analyzed, the black residue was found to be charcoal debris from free carbon. Some of that material was identifiable to a type of shrub. Several years later, Pei found additional burned bone fragments, quartz, and other stone artifacts. Hominin remains were also found, although not directly associated with the tools. These were two adult *Homo erectus* jaws and a large skull fragment. The human material was dated to 500,000 years old. Scientists of the time readily assumed that *Homo erectus* had used fire in his daily life. Then additional studies suggested that this could not have been the case since it could not be ascertained whether the bones were burned in the caves while *Homo erectus* was there or if they had been burned elsewhere by a natural fire and later washed into the cave. The relationship of human bone to the fire was not sufficiently secure to give *Homo erectus* the edge. Some archaeologists considered the lack of *defined* hearths significant. They also analyzed the ash layers that consisted of owl pellets and hyena scats that had been burned, probably from spontaneous combustion or natural surface fires. Manganese can stain bone, making it appear burned, and may have done so at Zhoukoudian, confusing the issue still further. In 2004, however, a statistical study using three-dimensional computer analysis located each bit of material found. The resulting distribution map clearly suggests that Zhoukoudian was a living site, and its inhabitants apparently were using fire (Boaz et al. 2004), now dated at between 600 and 800 kya (Shen et al. 2001).

Chinese archaeology has altered the evolutionary landscape in the past several years, finding older and older material, enriching our knowledge of what happened in human history. At the site of Xihoudu, some thirty rudimentary stone tools were found in close association with remains of deer and horse, among other large mammals, that were discolored to black, gray, and gray-green, suggesting that these bones had been burned. Originally dated to 1 mya, paleomagnetic studies suggest an older date, back to 1.8 mya, contemporaneous with Olduvai Gorge in East Africa. While the bones were burned, the association with the tools is not certain. These tools may actually have been brought into the site by water transport, and not been associated with the bones found nearby, so the bones may not have been burned by human agency. Some scholars are adamant that there still is no acceptable evidence of hearths let alone fires before 400 kya (de Lumley 2006, 150):

> Dans tous ces sites très anciens, antérieurs à 400,000 ans, aucune preuve évidente témoignant de l'existence de foyer aménagé, volontairement allumé et entretenu, ne peut être mise en évidence jusqu'à present.
> *In all these sites dating from before 400 kya, there is neither proof nor evidence until the present that attests to a dedicated hearth that was purposefully lit and maintained.* [Translation mine]

Human use of fire in Asia, as elsewhere, clearly increased by 400 kya. By around 300 kya, fire is well attested to in Western Europe and in China under glacial conditions. Inhabited caves have dugout and paved hearths, considered by archaeologists to have been usable both as cooking and roasting pits. Interestingly, like Koobi Fora over a million years earlier, a hearth was found at the mouth of a cave in the Dordogne, France (dating from about 100 kya), interpreted to have been there to keep carnivores out, as well as to provide warmth and light (Straus 1989). This is the use to which fire may have been put originally (Bellomo 1994a, 1994b). From this time period the evidence of hearths is unassailable. And what's more, there is evidence of the "home" in which the hearth became the heart. At Lazaret, in Nice in southern France, and dating from the fossil known as archaic *Homo sapiens* at between 190 to 160 kya (de Lumley 2006), is a structure built against a cave wall. The earlier the site, the more likely it is to be in the open; cave sites became popular with *Homo erectus* and the australopiths. The one at Lazaret stood approximately 11 m (36 ft) by 3.35 m (11 ft). A framework of poles held skins that, like a teepee, served as the walls of this shelter. Bones and rocks supplied support at the base of the structure, and inside the shelter were two hearths that had once burned oak and boxwood. These woods apparently produce embers that are easy to rekindle (de Lumley 1976). But these sites are modern in comparison to the time period of the Ancestor.

I see the Ancestor at the first stage of domestication of fire, when associating with it and using it were also becoming habitual. They would have been doing what has been recorded for former hunters and gatherers: collecting a burning branch in a tree that had been set alight by lightning, they walk, carrying a firestick, brushing it up against grasses as they walk through a savanna valley, back to a wooded area. There they find the family group—several adult females, some adolescent children of both sexes, some juveniles, and some infants in arms. On the way, each female and her young gather fruits, berries, lizards, grubs, larvae, and nuts. They are camping here, near trees in which they may sleep. The large male will probably make a nest on the ground, as do contemporary adult male gorillas, or in the trees, like chimps, and perhaps a juvenile will sleep with him. He goes to an old, dry termite mound or tree stump and holds his firestick to it as twilight grows. It ignites, and the other males of the group, females, and their young cluster around. The juveniles play, throwing nuts at each other. Some fall into the fire and pop. Someone retrieves the opened nut with a stick and burns his fingers touching it. Thrown to the side it cools and someone else picks it up, takes it away a little distance, and eats it. For years the hair on their bodies has not grown as thickly. Without the fur covering, they feel cool at night; at dusk, the insects are troublesome. The warmth of the fire is pleasant. Insects seem not to come as close and certainly whatever-is-in-the-bush is afraid to approach. The light gives time to sit around, to prepare twigs for tomorrow's termite gathering, to feel warm and safe as they prepare for sleep.

The Spark That Ignited Human Evolution

I have looked at the evidence for fire and found that the earliest date for the acknowledged use of fire by our ancestors is nearly 2 million years old (give or take 200,000 years). Is my extension back toward 6 million years ago for first contact with fire warranted? Let's take a look at the evidence. The review of chimp and monkey behavior indicates that the brain's ability was there. Fossil evidence gives the locomotor and postural evidence of a habitual biped. Geological information gives a habitat of woodland that is increasingly becoming savannalike. Genetic evidence notes the differences between chimps and humans and places these differences back toward this early date, when bipedalism was first established. The genetic evidence further notes that the kinds of differences are not only in the body, but also, and significantly, in the brain. Inferences drawn from postural evidence and genetics suggest that hairlessness began somewhere in this period as well, a function of variations in a particular gene (Rogers et al. 2004), which was in itself a corollary of bipedality and living in increasingly open areas, exposed to more heat than there is in the forest.

Anecdotes about, observations of, and experiments with modern apes (e.g., Chamove 1996) indicate that all apes will approach, or even play with, fire. Dietary evidence gives the motivation for doing so: insects that approach fire will be as tasty as termites themselves, and not only apes eat termites, but australopiths continued to do so after the 4-million-year mark (Sponheimer et al. 2005). I conclude therefore that motivation, ability, circumstance, and environment merged to inaugurate this unique relationship to fire beginning around 6 mya.

My view of human evolution is that the acquisition of fire was the engine that propelled the incredibly fast evolution of humans. Directly or indirectly, it affected cognitive processes, social processes, genetic systems, reproduction, the immune system, and digestion, among others. It may even have enhanced hair loss. The speed is reckoned from a geological perspective. The hominin trajectory is incredibly fast: beginning more or less at around 15 to 19 million years ago, and by 6 million years, hominins are moderately bipedal. By 4 million years they are advancing rapidly in facial and dental structure, and by 2 to 2.6 million years we have evidence of people using stone tools, taking the direction of their own change into their own hands. By just over 1.5 million years ago, our ancestors—now a "cosmopolitan" species found in Africa, Europe, and Asia—were leaving traces of fire use. This is a qualitatively different trajectory from other primates, who were evolving over the same time span. The pattern of change, the rate of change, and the number of genes involved are apparently unique to our lineage. Bruce Lahn and his research teams have demonstrated changes in particular genes that regulate development of the cerebral cortex (Evans et al. 2004). They and others (Kaessmann and Burki 2004) have shown that there are at least 214 genes that have undergone rapid change over the recent tens of millions of years, the time covering the divergence and evolution of humanity. And these are *our* changes—both the result of and the cause of further development toward the human line.

The mapping of the genomes of a variety of mammals and humans has demonstrated that the evolution of proteins related to hominin evolution is considerably faster in primates than in other mammals (rodents, for example), and the speed is particularly remarkable in the genes related to nervous system development (Dorus et al. 2004). The protein building block of naturally occurring opioids unique to humans is a further illustration of both rapid evolution in the human line and *different* emphasis in brain evolution (Rockman et al. 2005). The increasing adaptive value of changes in these endorphins involving emotion, fear, pain, and learning is one more indication of how mind and mental processes were changing in hominins at the divergence. The selective pressure was intense and directional, emphasizing nervous system change over other developments. Behavior (social, reproductive, cognitive) is the *emergent* consequence. In turn, the expansion of the behavioral repertoire itself selected for further cerebral enhancement and led to a continuous feedback of more changes.

It is hard to identify exactly *why* the Ancestor first *approached* fire. It surely must have occurred, however, somewhere between 7 and 10 million years ago, as apes were becoming more like contemporary chimpanzees and diverged from what became humans. That step was literally a step, as it depended on the origin of bipedalism. Old World monkeys sit erect, using their hands as tools to manipulate and manufacture. Freedom of the forelimb preceded committed bipedalism. As an outgrowth of sitting erect, this freedom enhanced, and perhaps encouraged, the development of committed bipedality. When you watch a cat or dog, or even a raccoon or squirrel, trying to get something, open something, or even scratch, you come to appreciate how vital an organ the hand really is, with its mobile digits and opposable thumb.

The lumbar lordosis, that s-shaped curve in the lower back that is so vital to erect posture and bipedal locomotion, develops in the human infant and toddler as a function of sitting and crawling. The inclination or propensity to develop that bend in the lower back is genetic, but the actual development occurs within the growth period as a function of activity. The bipedal shuffle of monkeys and apes—whether entrained or developed occasionally in the wild—is about as far as they can go. The shuffle, however, as exhibited especially by bonobos, is a better candidate for the immediate predecessor of bipedalism than is the knuckle-walking gait of contemporary African apes, which requires incredible development of the neck and shoulder muscles, as well as specific adaptations in the wrists (Dainton 2001).

Bipedalism means that the hands are free to do something other than carry body weight. It also implies a different relationship to heat as the vertical posture provides more surface exposure to breezes, an especially important consideration for a creature becoming increasingly hairless. Loss of hair and pigmentation accompanied this process (Rogers et al. 2004), abetted by the association with fire since naked skin absorbs more energy in the heat of the day, but also loses more heat during the night. Does it follow that without a thick coat the hominin would have felt the sting and bite of insects more than before? Or the chill of night? Living in woodland-grassland habitats that have more open spaces than do tropical forests, the Ancestor would have experienced yearly temperature ranges that currently are from 15°C to 30°C (60–85°F) (Gommes 2002; Alsop 2007). This would be balmy—in the *daytime*—but a nighttime drop to even 20°C (77°F) would seem cool, and fire might be nice to take away the chill.

The Ancestor was a being as competent to live in the mosaic of woodland-savanna as a baboon or macaque. In chapter 5 I talked about the cognitive and social abilities of monkeys and apes, which are far more subtle and complex than was accepted even 30 years ago. These nonhuman primates have culture, developed from traditions, which are passed on and altered generation to generation. They think, have "mind," and have neurons that give evidence of their ability to appreciate and act in terms of what they know the other guy knows. There is also evidence that the sounds and gestures that apes, and even monkeys, make carry more information than simple concepts like "danger" or "food." Monkey calls are now known to have *sequences* that identify who, which way, and what (Arnold and Zuberbühler 2008). Regular, agreed-upon sequences suggest a proto-grammar (Arnold and Zuberbühler 2008). Not only must the primates' knowledge of where and when food becomes ripe have to be memorized or analyzed, but because they live in social groups, enormous mental energy is required in this interpersonal domain as well. It takes a fair degree of skill to negotiate relationships and to predict the actions of others as well as the environment. Remembering, deceiving, cooperating, and associating memories to act in new ways are skills necessary in a society (Dunbar 1996). Primates are social creatures and need to keep track of who is who in their group, who is related to whom, who is apt to behave this or that way, or who is entitled to such and such because of status or rank. The list goes on.

It has taken a strangely long time to acknowledge those abilities that were and are still necessary for nonhuman primates to live in their habitats. The Ancestor had at least these monkey capabilities, as well as those of the African apes, especially the bonobos, with whom we share. so much genetic material. And the evidence is there to show that chimps of both kinds (common and bonobo) think and, like monkeys, navigate through difficult social terrain, have culture and some sense of numbers, create tools specific to a particular job, and plan for future actions. Monkeys and apes hunt, although the chimps do so in a more concerted, organized fashion than do monkeys. Hunting is serious predation. By definition, predation requires forethought, sustained intensity, organization, and a system of distribution of the kill (Teleki 1981). These requirements are all met by the behavior of the common chimp. While bonobos also hunt, their activity seems more ad hoc than that of their cousin (Hohmann and Fruth 2008). Interestingly, neither species kill their prey first—something that true predators do.

This competence of those on the "path not taken" reminds us of the enormity of the differences between us. At the same time, however, accepting our similarity enables us to conjure up an image of the Ancestor. The evidence for committed bipedalism in the human line starts with *Orrorin tugenensis* at 6 mya and is indisputably clear at 4 mya. While the gait certainly cannot have been like ours—due to major differences in muscles, bone, and spinal nerves—it was a gait redefining the human relationship to its environment. The head of *O. tugenensis* was up, eyes forward, with overlapping fields of vision giving good depth perception, neck turning easily and improving peripheral views, with hands free to carry, hold, and retain something desired for another time. Grant this person chimpanzee and bonobo abilities, plus some. The plus comes from studies of the differences between chimp and human brains. The genes controlling cortical brain development have altered at the rate of one advantageous change every 300,000 to 400,000 years (Dorus et al. 2004; Evans et al. 2004). This would have permitted at least eighteen to twenty changes in the cortical infrastructure between the time of the divergence and the appearance of *Orrorin*, allowing further human development in mind and brain. That means that the basic chimplike brain, despite its small size, had already developed capabilities that permitted new cognitive skills. These changes help to make my scenario plausible. Here is a being capable of working the environment as no creature has done before, and in so doing, govern its own evolutionary trajectory. The key factor is the relationship of the Ancestor to fire. The following model of fire acquisition was presented in chapter 2.

1. Approach and associate
   a. food
   b. protection (warmth, light)

2. Nurture
   a. dried grass
   b. twigs

3. Manufacture—the invention of *Homo sapiens* (sometime after 500 kya)

The African apes either do not have, or easily overcome, fear of fire. Even the Asian orangutan is known to imitate people in setting things alight. Gorillas will approach fire, and Chamove's experiment with captive wild-born chimpanzees (chapter 2) showed that they displayed increasingly less excitement toward fire over the course of several days, to the point of indifference. They certainly did not exhibit fear. This is powerful evidence. In addition, we know that monkeys and apes adore termites and other insects, and that fire-loving insects approach fire-burned areas. I speculate that, like birds of prey that fly *toward* a fire front, looking for animals running before it, the Ancestor would have approached areas where fire was burning to await the arrival of dinner. Knowing that young chimps play in imitation of their elders as well as by innovation, it is not beyond possibility to speculate that a juvenile would pick up a burning stick and, like the captive chimps, fling it around in play.

One way traditions become part of the behavioral repertoire is simply by juvenile innovations becoming included by dint of that innovator's repeated use of the pattern. By the time the juvenile is an adult member of the group, and assuming that this particular juvenile has sufficient prominence within the group, all younger animals will imitate what appears to them to be established group behavior. Hence, picking up burning sticks might quite readily have become part of what they did. The more experience, the larger the repertoire on which to base future behavior, the more memories to share, and the more new patterns possible. Over the years—and we have millions of them to consider—palpable comfort, habit, and the iterations of pattern on pattern would most plausibly have made the Ancestor a committed user of fire, eventually seeking it not only for its association with insects, but for protection, heat, and light. Settling near a burning stump requires nothing but willingness to be close to it. The next step would be to realize that the fire would last longer if twigs or branches were added as the stump burned down. Even contemporary "traditional" peoples, with sophisticated knowledge of how to make fire, were known to prefer the easier route of tending natural fires, feeding them bits of twig and dried material, and keeping them going for considerable lengths of time.

Protection may well have been an important motivator for choosing a site near a source of fire. Pick up a burning branch (behavior stemming from play behavior), brandish it at a potential predator, and the little hominin becomes a giant among its foes. The Ancestor was introduced as a small being, the prey of the big cats roaming freely through open forest, woodland, and savanna (Hart and Susman 2005), as chimps are to lions (Tsukahara 2005). Birds of prey also attack primates, and chimps are known to attack large carnivores in turn (Hiraiwa-Hasegawa et al. 1986). In 2006 scientists discovered that the famous Taung child, found in South Africa in 1925—and the first australopith to be identified as a human fossil—had died from an eagle attack (Berger and Clarke 1995; McGraw et al. 2006), not unlike the way juvenile macaques in Hong Kong are subject to "scalping" by black-eared kites.

The Ancestor may have been small, yes, but equipped with a mind quite capable of understanding that fire kept the beasts at bay, which was critical in order to survive the dangerous conditions in open areas. Other attributes of fire would have enhanced the Ancestor's relationship to it. Heat, as a key example, may well have enhanced the relaxation of selective pressure, eventually rendering the ancestor hairless. Furthermore, the ancestor lived during a period of rapid change (geologically speaking). Over those 3 million years between 7 and 10 mya, humans were diverging from apes and finally speciated somewhat less than 6 million years ago (Patterson et al. 2006). The climatological cooling in the Early Miocene encouraged the growth of woodlands over tropical forests. By the time of bipedal *Orrorin tugenensis* in the Late Miocene, there were the beginnings of a wet, warming trend, with a dry season that lasted about 3 months (Reed and Fish 2005) and provided more food from plants like grasses, their seeds, stalks, and rootlets. A new cornucopia of foods became available and these contained substances to which the evolving hominin had to adapt as these plant substances interacted with their physiology. But what did they eat? Primate diets are known from observation. The majority of nonhuman primates, including the great apes, are able to exploit diverse diets so that even serious hardships like droughts can be handled. As less food (and therefore energy) becomes available in the dry season, the nonhuman primate response is to become less active and/or to change group size to facilitate foraging (either getting larger to cover more ground and inform each other or smaller to compete less for limited resources) (Brockman 2005).

Altering diets can be dangerous. Plants contain a variety of secondary compounds that serve to protect them from insects and even mammalian predators (Stahl 1984). Toxicity of these compounds varies, but where preferred foods are scarce, monkeys (and other animals) become geophagous. They have been seen to eat charcoal (Struhsaker 1997) and, more typically, to eat soil, which either neutralize or absorb substances that would make them ill (Burton et al. 1999). Yet a major concomitant to bipedalism was an apparent alteration in diet over the millions of years between the divergence and *Homo erectus*. Meat became an increasingly important adjunct, providing high-quality nutrients all in one package. There is a horse-and-cart issue here since recent observations confirm that, between 3 and 5 years of age, if the diet is high in animal protein, menarche tends to be earlier (Parent et al. 2003). Fire and firelight provided the stimulus to alter reproductive cycles, which were enhanced by this change in diet, which itself was enhanced by fire.

Over time, hominins became more adept at capturing larger and larger animals. Bipedalism accompanies this, the proof being the fact that the intestinal tract of even contemporary humans is midway between herbivorous and carnivorous nonhuman primates (Milton 1993, 1999, 2003). When the genetic ability to synthesize vitamins C and B12 was lost, primates had to ensure a diet that supplied these required nutrients. Vitamin B12 enhances the action of melatonin (Ikeda et al. 1998; Hashimoto et al. 1996) and is either retained in the body by symbiotic bacteria or derived from animal sources, including eggs. Primates once manufactured B12, with the help of bacteria in the gut, or obtained it from insects. Termites, for example have a plentiful supply (Wakayama et al. 1984). Vitamin C, on the other hand, generally comes from fruits and vegetables, although a little can be obtained from animal meat as well. Variation in emphasis of diet may qualify certain species to be considered specialists. But overall, nonhuman primates eat from the same larder—they have the ability to ingest nutrients from the same wide variety of substances.

Early hominins also differed anatomically from their earlier cousins in that their molar teeth had thicker enamel, the jawbone was more robust, the molars were lower, and the incisors wider. Pitting on the incisors as well as striations on the molars indicate that abrasive vegetation was being used. The gritty material would have been found closer to the ground—that is, the hominins were not feeding from the trees as their more apelike ancestors had done. These microwear patterns on hominin teeth clearly indicate that the diet was coming from open areas. Living at a time when C4 plant availability increased, the human ancestor ate an eclectic diet found in woodland or savanna, not forest (Teaford et al. 2000; Scott et al. 2005). In these woodland-grassland areas, daytime temperatures drop, and fire at night would have been welcome. Our Ancestor might simply have sought out lightning-initiated, burning tree stumps or burning termite nests. Where available, in South Africa, for example, the increasingly hairless Ancestor may have gone in search of coal and peat burns. No intelligence or insight is required other than searching out burning stumps (from termite nests or trees), or coal seams.

The more accustomed the Ancestor became to the benefits of fire, the greater the dependence. It's the old "cell phone syndrome"—what began as a luxury became a necessity. Life is now virtually unlivable without it. And the dependence would have been real, by which I mean that things the Ancestor could do because of fire were essential to survival. Of these, a crucial asset would have been defense from predators. Unlike the Ancestor, the predators, snakes, and even insects that tormented humans would have kept a healthy distance from the smoke and fire of a burning stump—no small advantage in that. Certainly it would have meant that fire could not be taken for granted. It was too precious. As with the chimp experiment with fire, the Ancestor would have begun to feed it, and tending fire would have become part of the daily repertoire. The consequence of being near fire on a regular basis is central to my hypothesis.

Evolution implies flexibility. Nothing exemplifies this better than the circadian clock. This free-running rhythm that tends to flow around a 24-hour period is easily and readily—and evolutionarily—"intended" to entrain, to link in rhythm with some environmental cue. The rhythm is endogenous. It is, however, highly susceptible to both external influences, like light or behavior, and internal influences, like a secretion. Entrainment starts in the retina where light is first perceived. Photoreceptors in the retina transmit to the pineal gland via the suprachiasmatic nucleus of the hypothalamus in the brain, thereby driving mammalian rhythms from sleep, body temperature, and hormonal levels, to cognitive performance, eating, and reproduction. The arrival of the transduced light signal releases melatonin via hormones, neurons, and cerebrospinal fluid, directly into the blood stream and readily conveys the photoperiod to every tissue and organ, synchronizing daily and annual rhythms (Pang et al. 1998).

Melatonin, as a soluble fat, is not subject to barriers that impede other substances from circulating throughout the body. It can therefore move into the brain, which other substances cannot do. Its secretion is a function of the organism's response to day-night and light-dark cycles, increasing soon after darkness peaks and then gradually falling during the second half of the night. Even modern humans left in complete darkness for some time will show an approximately 24-hour, day-night activity cycle. The associating of this rhythm to environmental and social cues is the point. The trigger may be direct, as light itself, or indirect, through hormones in plants (Murch et al. 2001). Reviews of conception and birth rates throughout the world pegged to latitude (and therefore the amount of light over the course of the year) are inconsistent. Periodicity in humans, while it exists, is not strongly marked in the constant day length at the equator, according to some scholars (Roennenberg, 2004). Others suggest that independent of latitude, there "is conflict" in the data concerning seasonal variation (Bronson, 2004), and that perhaps because of indoor lighting, clear effects of seasonality do not show up.

It is this effect of light, interrupting patterns of seasonality, which is important to the issue here. People in the Arctic have fewer conceptions in the winter months, when darkness prevails for most hours of the day. Under conditions of darkness during these winter months, melatonin is amply secreted in the pineal gland (Brzezinski 1997) and conception is decreased. Duration of nocturnal melatonin secretion is directly proportional to the length of the night. It is the *suppression* of melatonin by light—the increase in "daytime"—that causes the cascade of sexuality and fertility. Dopamine functions in daylight. In an evolutionary sense, then, the suppression of melatonin secretion by extending the activity of dopamine reinforced the tendency to pair bond (Young and Wang 2004) while lengthening the time of learning, as, for example, in story-telling and direct instruction.

Since the effects of melatonin are greatest 2 hours after the onset of darkness (Goldman, 2001) the firelight would have mitigated melatonin's effect. Suppression of melatonin means that the usual target organs are not being stimulated. The onset of true darkness occurred only when (or *if*) the campfire was left to go out during sleep. Melatonin ceases to "flow" at dawn and dusk (Danilenko et al. 2000). During the first 6.5 hours of the night, humans are particularly responsive to light suppressing melatonin, and the consequence is a shift in the circadian clock (Zeitzer et al. 2000). Evening light or rather the absence of it, is essential to circadian patterns. Rather than being exposed to melatonin for 12 hours (e.g., 5:00 p.m. to 5:00 a.m.), as is the case with dawn-to-dusk mammals, in the presence of firelight, duration of melatonin secretion would have come closer to 8 or 9 hours, or perhaps even less (e.g., 8:00 or 9:00 p.m. to dawn at 5:00 a.m.). A 50 percent decrease in exposure, due to the decrease in hours of greatest impact, would have had repercussions throughout the system's melatonin affects. Melatonin is a powerful antioxidant, and this benefit is considered a reason that sleep exists so universally among vertebrates (Hobson 2005). A lack of sleep might even be a cause for cancer (Raloff 1998), given melatonin's role.

The nonvisual system is most sensitive to blue light at 446–77 nanometers and can phase shift—that is, reset the ordinary schedule in modern urban subjects under experimental conditions—by double the effect of green light (Lockley et al. 2003). Campfire light provides blue and green in the flames closest to the ground, and white and yellow at the top. Studies on seasonal affective disorder have demonstrated that these bright lights, for example 2,000 lux, make an impact on the disorder, introducing sufficient light to stimulate the suppression of melatonin (Hatonen et al. 1999).

Sometime in our evolution a process began that dampened the effects of the environment on human physiology. In geological time, as mountains rose, seas shrank, magnetic poles shifted, and salinity of the oceans changed, seasonality became more marked. That is, changes in the lengths of four-season patterns in temperate climates, and wet-dry alternations in tropical zones, became more distinct. Somewhere in the Late to Middle Miocene, for example, woodlands become more seasonal in East Africa (Andrews and Kelley 2007). The most significant increase occurred after 3 mya (Reed and Fish 2005). As seasonality became more and more pronounced, hominins and their descendants found ways to enfranchise themselves from it. Anthropologist D. K. Brockman suggests in the following quote (2005, 543) how hominins profited from seasonality:

> [I]t appears that hominins, or at least some of them, took advantage of increasing seasonality to thrive at the expense of other primates, and the novel way of dealing with seasonality was a critical part of the hominin lineage.

What could have been the novel way of dealing with seasonality suggested by Brockman? The thesis of this book is that fire was that novelty, making cold seasons warmer and nights shorter—in effect, creating microclimates amenable to the Ancestor's way of life. The end result was that firelight, in extending daytime, shifted circadian rhythms. Patterns of seasonality in contemporary humans are all over the map. There is no clear picture, as might well be expected. The millions of years of exposure to light—and the increasing intensity and duration of light over millennia—did inexorably alter light-based patterns. Borchert and his colleagues have recently demonstrated that reproduction in plants is set off by the variation in light even at 1 degree from the equator, and sunset seems to be the trigger (Borchert et al. 2005). The connection with primate reproduction begins with the fact that seasonal primates are said to be dependent on food availability as the trigger to reproductive rhythms, and plants produce phytoestrogens (Parent et al. 2003). The amount of it varies depending on plant type, and because it slows or even stops reproduction, it is being considered as a pesticide against insects (Kolar et al. 2005). Melatonin as well as serotonin is found in plants (Murch et al. 2001). So through ingestion, nonhuman primates will receive signals that affect their reproductive states. Pang and his colleagues (Pang et al. 1998) recognized that because melatonin acts on several sites involved with reproduction, the effects add up, thereby influencing the photoperiodic control on reproduction. Since animals, including humans, use photoperiod to regulate reproduction, this summative effect of melatonin has a direct control on reproduction.

Descriptions of life histories in nonhuman primates indicate that this rhythmicity corresponds to plant cycles. Several things about this are interesting. For instance, plants contain a variety of hormones, as do animals, which control growth, reproduction, and just about everything else. Plant cycles are, of course, entrained to sunlight, which, due to the rotation of the earth around the sun, dictates seasonality. Melatonin secretion is highest 2 hours after darkness (Goldman 2001). Whether it is the timing of light exposure (Goldman 2001) that is important, or the duration of melatonin secretion through the dark part of the 24-hour cycle that makes the difference, is still controversial, but the decrease in length of night would seriously affect the brain's interpretation of seasonality, creating long days of short ones by virtue of firelight. Less debatable is the wavelength that triggers the cessation of melatonin secretion.

If melatonin is not available because it is suppressed by light, gonadotropin-releasing hormone (GnRH) is free to activate the gonads. The amygdala is affected by gonadal steroid hormones at puberty. The behavior of adolescents is qualitatively different from that of juveniles because of the many changes that take place in risk taking, planning, reward incentive, and even decision making (Sisk et al. 2004) during this time. The result is "new" persons with different perceptions of danger compared to the juveniles they were or the adults around them. As these adolescents mature, they would have memories of their own juvenile experiences, and their memories might take precedence over ancient fears. If parents are afraid of fire but their children play with burning twigs, those juveniles will grow up to continue using burning twigs without developing the fear their parents felt. Studies on how short-term memory becomes long-term memory indicate an epigenetic route (Colvis et al. 2005) and suggest that this remembering of behavioral innovations would affect subsequent behavior in evolutionarily significant ways. Traditions develop in this way just by moving up the age pyramid. As the young in turn become adults, the younger ones of the new generation absorb and imitate what they perceive as patterns that have always been practiced.

But was sitting around burning fires enough? I speculate on the role of fire because recent studies indicate that the human system does not work exactly like that of other mammals. It *could* do so, in that the endogenous free-running clock of human beings describes a nearly 24-hour periodicity, but the expression of the clock is highly variable. Of course, it is highly likely that some of this contemporary variability is a function of human evolution itself over the past several hundreds of thousands of years. We know that "genetic load" and release from some genetic constraints is a function of life after the advent of plant and animal domestication, some nearly 20,000 years ago. If we assume one generation every 20 years over a period of even just a million years, this still represents fifty thousand generations—a considerable time for evolution to accumulate changes, some of which are self- (human) propelled. Recall that advantageous mutations in the genes that control cortical growth of the brain occurred between 300,000 and 400,000 years ago (Evans et al. 2004).

Recent studies evaluating whether or not there has been a decrease in the age of onset of puberty are also confounded by socioeconomic variables related to diet, temperature, and general well-being. Furthermore, environmental contaminants in the form of pesticides and other endocrine-disrupting chemicals are considered to play a role in pubertal trends in modern populations (Parent et al. 2003). Given these modern realities, the reconstruction of events in hominization that took place in the millions of years ago may be futile, but it is certainly based on how mammalian systems work and what is still true in nonhuman primates (and perhaps humans in general).

It is tempting to think that the inconsistencies in human response to light and melatonin are a reflection (pun intended) of our very long history with artificial light, the heterogeneity of pattern being the result of relaxation on the genetic system (Goldman 2001; Wehr 2001; Roennenberg et al. 2003). While the lighting invented during the Industrial Revolution may have intensified and accelerated change, the original drive occurred at a far earlier date.

What is not clear-cut about my speculation concerns the manner in which the Ancestor would have had to stare at the fire, specifically gaze, timing, intensity, and duration. Studies on seasonal affective disorder are contradictory with reference to where the eyes must be looking. In the 1990s it seemed that illumination directed at peripheral vision was just as effective as that directed straight on (Adler et al. 1992). A decade later, however, studies were suggesting that it is the nasal portion of the retina that responds most (Ruger et al. 2005). Hence, it was thought that the eyes of the person must be oriented in the direction of the fire and must be open in order for the light to make a difference. Furthermore, the effect of light drops quickly according to the inverse square law, where the intensity of radiation is inversely proportional to the square of the distance from its source, so that a doubling of the distance from the fire will reduce the intensity of the energy by a fourth. Therefore, it would seem that for the campfire light to have had an effect, the group must have been rather close to it. Yet in the past 2 years, research has shown that light coming into the retina from above hits the inferior retina, and this is more effective in suppressing melatonin than when it hits the nasal portion of the retina. The inferior retina is therefore the more important contributor to light-induced suppression of melatonin (Glickman et al. 2003), which indicates that a person could be lying down and still be affected, as could happen in cozying up near to the fire as nighttime falls.

I reviewed an online database on ethnographic descriptions (the Human Relations Area Files) to find out where people were in relation to fire, and found that, of course, there was considerable variation depending on the nature of the shelters they built. Hunters and gatherers, however, who model the condition of the Ancestor, although they are far more sophisticated in their technology, knowledge, and adaptations, built cursory shelters, compatible with a nomadic life. Fire was a major source of heat for those shelters, such as are built by desert dwellers in the Western Desert of Australia, the Kalahari Desert of southern Africa, or even the Guayaki (Ache) of Paraguay, who live in "thick, tropical forest."

The following description of the Guayaki (Clastres 1972, 146) is quite suggestive:

> The fire is situated at the edge of the hut, just within the limits of the roof. Three or four large logs feed the fireplace, and at night the whole family lies on the ground in the little space left between the basket and the fire. The children nestle against their parents, facing the flame, with the adults behind them. As the heat diminishes, the father or mother gets up and replenishes the logs. Since the nights are often cool, each one in his sleep tends to draw nearer the fire, and the morning reveals a pile of naked bodies half covered with ashes.

The amount of light will not have exceeded 20 lux, but the duration would have been well into the night. Studies have shown that even a modest 10 lux is sufficient to shift melatonin rhythms (Ruger et al. 2005), although stronger illumination produces stronger responses. Researchers have repeatedly been surprised at how little light is required to affect the system. The lowest figure I have seen is 1.3 lux (Raloff 1998). I searched the Human Relations Area Files to find out when peoples before the "present" (i.e., 1965) went to bed. And although cultural anthropologists favor the description of kinship systems over daily life, I was able to find a few descriptions that affirm that people did not go to bed as the sky darkened—contrary to what a few neuroscientists have conjectured (e.g., Wehr 2001; Bronson 2004). The "dawn to dusk" image of the daily habits of nonurbanized people is simply not how it happened. Whether the group was agricultural, like the Azande of Central Africa, the Ibo of Nigeria, the Zulu of South Africa, or hunters and gatherers, like the San of the Kalahari Desert or the Paliyan of Pakistan, they were either minimally flexible at bedtime hours or decidedly not restricted by sundown. The Baganda of Uganda are an interesting exception because while the ordinary folk went to bed at sundown and were up at sunrise, the chiefs had torches and stayed up till whenever. Rather than "lights out" at dusk like most mammals, the people in traditional times sat around the campfire, working on tools, or looking across the flames to the storyteller as he wove the tales of the people.

SEVEN

People generally do not like to move outside of the aura of light. Ghosts move around in darkness—nearly universally throughout the world's cultures. Where wild animals roamed, there was always the fear of their intrusion contrasted with the security of the flames. Nisa, a child of the San, put it this way:

> The lions came near our camp and stayed just outside the circle
> of firelight. We could see their eyes shining out of the darkness
> of the surrounding bush. One pair of eyes were in one place,
> another pair of eyes were in another place, and there were others.
> They were many, they wanted to kill us. (Shostak 1981, 76)

This example from a child of the San culture underscores the role fire came to play as a powerful source of protection from predators, such as lions. But did humanity also begin to expose itself to severe—relatively unknown—diseases at the onset of living with fire? Certainly the particulate matter in the smoke of a fire is a disease agent, among which carbon monoxide and methane are best known (Huttner et al. 2007; Platek et al. 2002). These substances and others provoke chronic obstructive pulmonary disease (COPD) or lung cancer. Serious injuries, especially to children, are inflicted by open flames, and if any meat fell into the fire (this is long before controlled cooking) and was eaten, the risk of developing cancer from heterocyclic amines increased substantially (Grover and Martin 2002). As the Ancestor moved into new habitats, encountering different food sources that would have ultimately slowed "transit time" through the gut, this anticancer attribute of melatonin would have been particularly important (Ragir et al. 2000).

The vertebrate body evolved with defenses against toxins. After all, plant material contains natural pesticides that are intended to keep predators away from the plant. The kidney and liver are the organs responsible for detoxifying, and what they do is extraordinary, but there are limits. In addition, the body refurbishes itself by sloughing off old cells, especially in the digestive system (Ames et al. 1987). Among the more powerful defense mechanisms is, of course, the flow of antioxidants, either taken from plant material (as are vitamin C or carotenoids) or secreted by organs—especially melatonin. Since light suppresses melatonin, firelight would shift the immune system, leaving some diseases free to develop, due to a loss of a component of the T-cell system (Nguyen et al. 2006). The extension of light into what should be darkness has an effect on stem cells and their manufacture of blood (Méndez-Ferrer et al. 2008).

So, while large mammals represent the precondition as they do not run from fire, apes suggest the first stage, able to approach and interact with fire, to pick up burning twigs and firebrands, and perhaps to play with them. The next step is where I place the Ancestor. With the abilities, attributes, and experience of those who came before, the Ancestor associated with and purposefully fed the fire. This is the first intimation of control and, with it, mastery of nature. It is the quantum leap after the divergence between our ancestors and apes. The initial control apparent in feeding the fire grows to include transporting it from hearth to hearth. The final step indicates control of the ability to manufacture fire from raw materials, along with the knowledge, technology, and cognitive skills to do this. This last step in the conquest of fire happened relatively recently in time, with a sapient form of *Homo sapiens*.

And there it is. This interplay of factors over millions of years took apelike beings who moved on two legs onto a trajectory that led ultimately to the beings we are today. With a brain no larger than a chimp's, but a mind accelerating genetically, anatomically, and behaviorally by leaps and bounds, these little creatures altered the face of the earth. They did so because they were more interested in delicacies found at a smoldering fire than they were afraid of heat or flame; they were playful and social and dared to pick up firebrands; they saw the advantage of brandishing these sticks at predators that came too close; and they found comfort in fire's warmth and perceived that putting twigs into it kept it going. They changed the timetable of growth, development, and reproduction because sitting by the fire altered the night's flow of melatonin and the cascade of hormones that follow it. Rates and patterns of growth, along with regulation and activation of genes, all changed as a consequence, accelerating processes of mind, body, disease, and society in the transformation to us.

References

Ackermann, Rebecca Rogers, and James M. Cheverud. 2004. Detecting genetic drift versus selection in human evolution. *PNAS* 101 (52): 17946–51.

Adler, J. S., D. F. Kripke, R. T. Loving, and S. L. Berga. 1992. Peripheral vision suppression of melatonin. *Journal of Pineal Research* 12 (2): 49–52.

Alexander, R. M. 1995. Standing, walking, and running. In *Gray's anatomy*, 38th ed., ed. P. L. Williams. New York: Churchill, Livingstone.

Alonso-Vale, Maria Isabel Cardoso, Sandra Andreotti, Sidney Barnabé Peres, Gabriel Forato Anhê, Cristina das Neves Borges-Silva, José Cipolla Neto et al. 2004. Melatonin enhances leptin expression by rat adipocytes in the presence of insulin. *American Journal of Physiology Endocrinology and Metabolism* 288 (4): e805–12.

Alsop, Z. 2007. Malaria returns to Kenya's highlands as temperatures rise. *Lancet* 370 (9591): 925–26.

Ambrose, Stan. 2001. Paleolithic technology and human evolution. *Science* 291 (5509): 1748–53.

Ames, Bruce N., Renae Magaw, and Lois Swirsky Gold. 1987. Ranking possible carcinogenic hazards. *Science* 236:271–80.

Andrews, P., and J. Kelley. 2007. Middle Miocene dispersals of apes. *Folia Primatologica* 78 (5–6): 328–43.

Anway, Matthew D., Andrea S. Cupp, Mehmet Uzumcu, and Michael K. Skinner. 2005. Epigenetic transgenerational actions of endocrine disruptors and male fertility. *Science* 308 (5727): 1466–69.

Aoki, H., N. Yamada, Y. Ozeki, H. Yamane, and N. Kato. 1998. Minimum light intensity required to suppress nocturnal melatonin concentration in human saliva. *Neuroscience Letter* 252 (2): 91–94.

Arnold, Kate, and Klaus Zuberbühler. 2008. Meaningful call combinations in a nonhuman primate. *Current Biology* 18(5): R202–3.

Asfaw, Berhane, W. Henry Gilbert, Yonas Beyene, William K. Harts, Paul R. Renne, Giday Wolde Gabriel et al. 2002. Remains of *Homo erectus* from Bouri, Middle Awash, Ethiopia. *Nature* 416:317–20.

Asouti, E., and P. Austin. 2005. Reconstructing woodland vegetation and its exploitation by past societies, based on the analysis and interpretation of archaeological wood charcoal micro-remains. *Environmental Archaeology* 10:1–18.

Aton, Sara J., Gene D. Block, Hajime Tei, Shin Yamazaki, and Erik D. Herzog. 2004. Plasticity of circadian behavior and the suprachiasmatic nucleus following exposure to non-24-hour light cycles. *Journal of Biological Rhythms* 19 (3): 198–207.

Backwell, Lucinda R., and Francesco d'Errico. 2001. Evidence of termite foraging by Swartkrans early hominids. *Proceedings of the National Academy of Sciences* 98 (4): 1358–63.

Badrian, N., A. Badrian, and R. L. Susman. 1981. Preliminary observations on the feeding behavior of *Pan paniscus* in the Lomako Forest of Central Zaire. *Primates* 22:173–81.

Badrian, N., and R. K. Malenky. 1984. Feeding ecology of *Pan paniscus* in the Lomako Forest, Zaire. In *The pygmy chimpanzee, Evolutionary biology and behavior*, ed. R. L. Susman, 275–99. New York: Plenum.

Banjjo, A. D., O. A. Lawall, and E. A. Songonuga. 2006. The nutritional value of fourteen species of edible insects in southwestern Nigeria. *African Journal of Biotechnology* 5 (3): 298–301.

Basden, G. T. n.d. *Igbo. FF26.* Human Relations Area Files. http://www. yale.edu/hraf/.

Bazar, Kimberly A., A. Joon Yunb, and Patrick Y. Leeb. 2004. Debunking a myth: Neurohormonal and vagal modulation of sleep centers, not redistribution of blood flow, may account for postprandial somnolence *Medical Hypotheses* 63 (5): 778–82.

Becker, Jill B. 2002. Introduction to behavioral endocrinology. In *Behavioral endocrinology*, 2nd ed., ed. J. B. Becker, S. M. Breedlove, David Crews, and Margaret M. McCarthy, 3–39. Cambridge, MA: MIT Press.

Bellingham, J., S. S. Chaurasia, Z. Melyan, Cuimei Liu, M. A. Cameron, E. Tarttelin et al. 2006. Evolution of melanopsin photoreceptors: Discovery and characterization of a new melanopsin in nonmammalian vertebrates. *PLoS Biol.* 4 (8): e254.

Bellomo, Randy V. 1993. Methodological approach for identifying archaeological evidence of fire resulting from human activities. *Journal of Archaeological Science* 20 (5): 525–53.

———. 1994a. Methods of determining early hominid behavioural activities associated with the controlled use of fire at FxJj20 Main, Koobi-Fora, Kenya. *Journal of Human Evolution* 27 (1–3): 173–95.

———. 1994b. Early Pleistocene fire technology in northern Kenya. In *Society, culture, and technology in Africa*, ed. S. Terry Childs. Philadelphia: University of Pennsylvania, Museum of Applied Science, Center for Archaeology.

Berger, L., and R. J. Clarke. 1995. Eagle involvement in accumulation of the Taung child fauna. *Journal of Human Evolution* 29:275–99.

Bermejo, M., and G. Illera. 1999. Tool-set for termite fishing and honey extraction by wild chimpanzees in the Lossi Forest, Congo. *Primates* 40:619–27.

Berson, D. M. 2003. Strange vision: Ganglion cells as circadian photoreceptors. *Trends in Neuroscience* 26 (6): 314–20.

Bickerton, Derek. 1990. *Language and species.* Chicago: University of Chicago Press.

Binford, L. R. 1983. *In pursuit of the past.* London: Thames and Hudson.

Bird, M. I. 1995. Fire, prehistoric humanity, and the environment. *Interdisciplinary Science Reviews* 20 (2): 141–54.

Bjorklund, D. F. 2006. Mother knows best: Epigenetic inheritance, maternal effects, and the evolution of human intelligence. *Developmental Review* 26 (2): 213–42.

Boaz, Noel T., Russell L. Ciochon, Qinqi Xu, and Jinyi Liu. 2004. Mapping and taphonomic analysis of the *Homo erectus* loci at Locality 1 Zhoukoudian, China. *Journal of Human Evolution* 46 (5): 519–49.

Boesch, Christophe, and H. Boesch. 1990. Tool use and tool making in wild chimpanzees. *Folia Primatologica* 54:86–99.

Boesch, Christophe, and Michael Tomasello. 1998. Chimpanzee and human cultures. *Current Anthropology* 39 (5): 591–614.

Bogin, B. 1998. From caveman cuisine to fast food: The evolution of human nutrition. *Growth Hormone & IGF Research* 8:79–86.

Bolton, K., F. D. Burton, and V. Campbell. 1998. Chemical analysis of soils of Kowloon (Hong Kong) eaten by hybrid macaques. *Journal of Chemical Ecology* 24 (2): 195–205.

Borchert, Rolf, Susanne S. Renner, Zoraida Calle, Diego Navarrete, Alan Tye, Laurent Gautier et al. 2005. Photoperiodic induction of synchronous flowering near the Equator. *Nature* 433:627–29.

Borjigin, Jimo, Xiaodong Li, and Solomon H. Snyder. 1999. The pineal gland and melatonin: Molecular and pharmacologic regulation. *Annual Review of Pharmacology and Toxicology* 39:53–65.

Bouton, M. E. 2002. Context, ambiguity, and unlearning: Sources of relapse after behavioral extinction. *Biological Psychiatry* 52:35–42.

Brain, C. K. 1981. *The hunters or the hunted? An introduction to African cave taphonomy.* Chicago: University of Chicago Press.

———. 1993a. Structure and stratigraphy of the Swartkrans cave in light of the new excavations. In *Swartkrans: A cave's chronicle of early man*, ed. C. K. Brain, 23–33. Pretoria: Transvaal Museum.

———. 1993b. A taphonomic overview of the Swartkrans fossil assemblages. In *Swartkrans: A cave's chronicle of early man*, ed. C. K. Brain, 256–64. Pretoria: Transvaal Museum.

Brain, C. K., and A. Sillen. 1988. Evidence from the Swartkrans cave for the earliest use of fire. *Nature* 336:464–66.

Brainard, G., J. Hanifin, J. Gresson, B. Byrne, G. Glickman, E. Gerner, and M. D. Rollag. 2001. Action spectrum for melatonin regulation in humans: Evidence for a novel circadian photoreceptor. *Journal of Neuroscience* 21(16): 6405–12.

Brewer, S. M., and C. W. McGrew. 1990. Chimpanzee use of a tool set to get honey. *Folia Primatologica* 54:100–104.

Brockman, Diane K. 2005. What do studies of seasonality in primates tell us about human evolution? In *Seasonality in primates: Studies of living and extinct human and nonhuman primates*, ed. D. K. Brockman and C.van Schaik, 543–71. Cambridge: Cambridge University Press.

Bronson, F. H. 2004. Are humans seasonally photoperiodic? *Journal of Biological Rhythms* 19 (3): 180–92.

Brosius, Jurgen. 2003. The contribution of RNAs and retroposition to evolutionary novelties. *Genetica* 118:99–116.

Brunet, Michel, Franck Guy, David Pilbeam, Hassane Mackaye Taisso, Andossa Likius, Djimdoumalbaye Ahountas et al. 2002. A new hominid from the Upper Miocene of Chad, Central Africa. *Nature* 418:145–51.

Brzezinski, Amnon. 1997. Melatonin in humans. *New England Journal of Medicine* 336 (3): 186–95.

Buijs, R. M., C. G. Van Eden, V. D. Goncharuk, and A. Kalsbeek. 2003. Circadian and seasonal rhythms. The biological clock tunes the organs of the body: Timing by hormones and the autonomic nervous system. *Journal of Endocrinology* 177:17–26.

Burling, R. 2005. *The talking ape: How language evolved.* New York: Oxford University Press.

Burton, F. D. 1972a. An analysis of the muscular limitation of opposability in seven species of Cercopithecinae. *American Journal of Physical Anthropology* 36 (2): 169–88.

———. 1972b. The integration of biology and behavior in the socialization of *Macaca sylvana* of Gibraltar. In *Primate behavior,* ed. F. Poirier, 29–62. New York: Random House.

———. 1977. Ethology and the development of sex and gender identity in nonhuman primates. *Acta Biotheoretica* 26 (1): 1–18.

———. 1984. Inferences of cognitive abilities in Old World monkeys. *Semiotica* 50 (1–2): 69–81.

Burton, F. D., and Mario J. A. Bick. 1972. A drift in time can define a deme: The concept of tradition drift. *Journal of Human Evolution* 1 (1): 53–59.

Burton, F. D., K. Bolton, and V. Campbell. 1999. Soil-eating behaviour of the hybrid macaques of Kowloon. *NAHSON Bulletin* 9 (4): 14–20.

Burton, F. D., and L. Chan. 1987. Interspecific babysitting. *Canadian Journal of Physical Anthropology* 65:752–55.

———. 1996. Behaviour of mixed species groups of macaques. In *Evolution and ecology of macaque societies,* ed. J. S. Fa and C. Southwick, 389–412. Oxford: Oxford University Press.

Burton, F. D., and L. Sawchuk. 1974. Demography of *Macaca sylvanus* of Gibraltar. *Primates* 15 (2): 271–78.

Byrne, Richard W. 2002. Emulation in apes: Verdict "not proven." *Developmental Science* 5 (1): 20–22.

Calarco, John A., Yi Xing, Mario Cáceres, Joseph P. Calarco, Xinshu Xiao, Qun Panet et al. 2007. Global analysis of alternative splicing differences between humans and chimpanzees. *Genes and Development* 21:2963–75.

Callinan, Pauline A., and Andrew P. Feinberg. 2006. The emerging science of epigenomics. *Human Molecular Genetics* 15 (Review Issue 1): R95–101.

CALnet. 2001. *Structure of the Retina*. Department of Anatomy, CALnet. http://137.222.110.150/calnet/visua12/page2.htm (accessed December 2004).

Cameron, Judy L. 1996. Regulation of reproductive hormone secretion in primates by short-term changes in nutrition. *Reviews of Reproduction* 1:117–26.

Cant, Michael A., and Rufus A. Johnstone. 2008. Reproductive conflict and the separation of reproductive generations in humans. *Proceedings of the National Academy of Sciences of the U. S. Biological Sciences/Anthropology* 105:5332–36.

Caro, T. M., D. W. Sellen, A. Parish, R. Frank, D. M. Brown, E. Voland, and M. Borgerhoff Mulder. 1995. Termination of reproduction in nonhuman and human female primates. *International Journal of Primatology* 16 (3): 205–20.

Carrillo-Vico, Antonio, Juan R. Calvo, Pedro Abreu, Patricia J. Lardone, Sofia Garcia-Mauriño, Russel J. Reiter, and Juan M. Guerrero. 2004. Evidence of melatonin synthesis by human lymphocytes and its physiological significance: Possible role as intracrine, autocrine, and/or paracrine substance. *FASEB Journal Express* 18:537–39.

Cashmore, A. 2003. Cryptochromes: Enabling plants and animals to determine circadian time. *Cell* 114 (5): 537–43.

Cerling, Thure E., John M. Harris, Bruce J. MacFadden, Meave G. Leakey, Jay Quade, Vera Eisenmann, and James R. Ehleringer. 1997. Global vegetation change through the Miocene/Pliocene boundary. *Nature* 389:153–58.

Chamove, Arnold S. 1996. Enrichment: Unpredictable ropes and fire. *Shape of Enrichment* 5 (2): 1–3.

Cheney, D. L., and R. M. Seyfarth. 1990. *How monkeys see the world: Inside the mind of another species*. Chicago: University of Chicago Press.

Cheremisinoff, Nicholas P. 1980. *Wood for energy production*. Energy Technology Series. Ann Arbor, MI: Ann Arbor Science Publishers.

Choctaw Tales. *Choctaw origin of fire*. 2005. http://www.ilhawaii.net/~stony/lore120.html.

Chomsky, Noam. 1953. Systems of syntactic analysis. *Journal of Symbolic Logic* 18 (3): 242–56.

Clark, Andrew G., Stephen Glanowski, Rasmus Nielsen, Paul D. Thomas, Anish Kejariwal, Melissa A. Todd et al. 2003. Inferring nonneutral evolution from human-chimp-mouse orthologous gene trios. *Science* 302 (5652): 1960–63.

Clark, J. D., and J. Harris. 1985. Fire and its roles in early hominid lifeways. *African Archaeological Review* 3:3–27.

Clastres, Pierre. 1972. The Guayaki. In *Hunters and gatherers today*, ed. M. G. Bicchieri, 138–74. New York: Holt, Rinehart.

Colvis, Christine M., Jonathan D. Pollock, Richard H. Goodman, Soren Impey, John Dunn, Gail Mandel et al. 2005. Epigenetic mechanisms and gene networks in the nervous system. *Journal of Neuroscience* 25 (45): 10379–89.

Committee on Animal Nutrition. 2003. *Nutrient requirements of nonhuman primates.* Washington, D.C.: The National Academies Press. http://books. nap.edu/openbook.php?record_id=9826&page=R1.

Cummings, D. E., K. E. Foster-Schubert, and Joost Overduin. 2005. Ghrelin and energy balance: Focus on current controversies. *Current Drug Targets* 6 (2): 153–69.

Curnoe, D., R. Grun, L. Taylor, and J. F. Thackeray. 2001. Direct ESR dating of a Pliocene hominin from Swartkrans. *Journal of Human Evolution* 40:379–91.

Czeisler C. A., T. L. Shanahan, E. B. Klerman, H. Martens, D. J. Brotman, J. S. Emens et al. 1995. Suppression of melatonin secretion in some blind patients by exposure to bright light. *New England Journal of Medicine* 332 (6): 6–11.

Dacey, Dennis M., Hsi-Wen Liao, Beth B. Peterson, Farrel R. Robinson, Vivianne C. Smith, Joel Pokorny et al. 2005. Melanopsin-expressing ganglion cells in primate retina signal colour and irradiance and project to the LGN. *Nature* 433:749–54.

Dainton, M. 2001. Palaeoanthropology: Did our ancestors knuckle-walk? *Nature* 410:324–25.

Danilenko, Konstantin V., Anna Wirz-Justice, Kurt Kräuchi, Jacob M. Weber, and Michael Terman. 2000. The human circadian pacemaker can see by the dawn's early light. *Journal of Biological Rhythms* 15 (5): 437–46.

Danthu, P., M. Ndongoc, M. Diaou, O. Thiam, A. Sarr, B. Dedhioud, and A. Ould Mohamed Valle Ould. 2003. Impact of bush fire on germination of some West African acacias. *Forest Ecology and Management* 173 (1–3): 1–10.

Dardente, Hugues, Jérôme S. Menet, Vincent-Joseph Poire, Dominique Streigcher, François Gauer, Berthe Vivien-Roels et al. 2003. Melatonin induces Cry1 expression in the *pars tuberalis* of the rat. *Molecular Brain Research* 114 (2): 101–6.

Dart, Raymond. 1925. *Australopithecus africanus:* The man-ape of South Africa. *Nature* 115:195–99.

———. 1957. The osteodontoderatic culture of *Australopithecus prometheus. Transvaal Museum Memoirs* 8.

Darwin, Charles. 1871. *Descent of man and selection in relation to sex.* London: Penguin.

de Heinzelin, Jean, Desmond Clark, Tim D. White, William Hart, Paul R. Renne et al. 1999. Environment and behavior of 2.5-million-year-old Bouri hominids. *Science* 284 (5414): 625–29.

de Lumley, H. 1976. Grotte du Lazaret. In *Sites paleolithiques de la region de Nice et grottes de Grimaldi,* 9th U.I.S.P.P. Meeting, Nice, France: Broché.

———. 2006. Il y a 400 000 ans: La domestication du feu, un formidable moteur d'hominisation. *Comptes Rendus Palévol* 5 (1–2): 149–54.

de Waal, F. B. M. 2007. *Chimpanzee politics: Power and sex among apes.* Baltimore: Johns Hopkins University Press.

Deaner, R. O., and M. L. Platt. 2003. Reflexive social attention in monkeys and humans. *Current Biology* 13 (18):1609–13.

DeFoliart, Gene. 1992. Insects as human food. *Crop Protection* 11:395–99.

Delagnes, A., and H. Roche. 2005. Late Pliocene hominid knapping skills: The case of Lokalalei 2C, West Turkana, Kenya. *Journal of Human Evolution* 48 (5): 435–72.

Dominguez-Rodrigo, M., J. Serrallonga, J. Juan-Tresserras, L. Alcalac, and L. Luquec. 2001. Woodworking activities by early humans: A plant residue analysis on Acheulian stone tools from Peninj (Tanzania). *Journal of Human Evolution* 40:289–99.

Dominy, Nathaniel J., and Brean W. Duncan. 2005. Seed-spitting primates and the conservation and dispersion of large-seeded trees. *International Journal of Primatology* 26 (3): 631–49.

Dohrn, M. 1995. *Crossroads of Nancite.* BBC Natural World. United Kingdom: BBC.

Dorus, Steve, Eric J. Vallender, Patrick D. Evans, Jeffrey R. Anderson, Sandra L. Gilbert, Michael Mahowald et al. 2004. Accelerated evolution of nervous system genes in the origin of *Homo sapiens. Cell Tissue Research* 119 (7): 1027–40.

Dunbar, R. I. M. 1996. *Grooming, gossip, and the evolution of language.* Cambridge, MA: Harvard University Press.

Electro-Optical Industries. *Educational and reference.* http://www. electro-optical.com/toplevel/educationref.htm (accessed 2003).

Engel, Cindy. 2002. *Wild health.* Boston: Houghton Mifflin.

Evans, Patrick D., Jeffrey R. Anderson, Eric J. Vallender, Sandra L. Gilbert, Christine M. Malcom, Steve Dorus, and Bruce T. Lahn. 2004. Adaptive evolution of *ASPM*, a major determinant of cerebral cortical size in humans. *Human Molecular Genetics* 13 (5): 489–94.

Evans-Pritchard, E. E. 1940. *The Nuer.* Oxford: Clarendon.

Falk, Dean. 1992. *Braindance.* New York: Henry Holt.

———. 2004. Prelinguistic evolution in early hominins: Whence motherese? *Behavioral and Brain Sciences* 27:491–503.

FAO (Food and Agriculture Organization of the U.N.) 2.1.1. 2001. *The fire environment, fire regimes, and the ecological role of fire in the region.* http://www.fao.org/documents/show_cdr.asp?url_file=/docrep/006/ ad653e/ad653e14.htm.

Farley, Kenneth A., David Vokrouhlicky, William F. Bottke, and David Nesvorny. 2006. A Late Miocene dust shower from the break-up of an asteroid in the main belt. *Nature* 439 (7074): 295–97.

Farm 215. 2006. *The fire of 2006.* Nature Retreat & Fynbos Reserve. http:// www.farm215.co.za/fire.html.

Fazzari, M. J., and J. M. Greally. 2004. Epigenomics: Beyond CpG islands. *Nature Reviews Genetics* 5:446–55.

Filler, A. G. 2007. Homeotic evolution in the mammalia: Diversification of therian axial seriation and the morphogenetic basis of human origins. *PLoS ONE* 2 (10): e1019.

Fischer, Tobias W., Andrzej Slominski, Desmond J. Tobin, and Ralf Paus. 2008. Melatonin and the hair follicle. *Journal of Pineal Research* 44 (1): 1–15.

Fischer, Tobias W., Trevor W. Sweatman, Igor Semak, Robert M. Sayre, Jacobo Worstman, and Andrzej Slominski. 2006. Constitutive and UV-induced metabolism of melatonin in keratinocytes and cell-free systems. *FASEB Journal* 20:1564–66.

Fisher, J. W., and H. C. Strickland. 1991. Dwellings and fireplaces: Keys to Efe Pygmy campsite structure. In *Ethnoarchaeological approaches to mobile campsites: Hunter-gatherer and pastoralist case studies,* ed. C. S. Gamble and W. A. Boismier. Ann Arbor, MI: International Monographs in Prehistory, Ethnoarchaeological Series 1.

Fondon, John W., and Harold R. Garner. 2004. Molecular origins of rapid and continuous morphological evolution. *Proceedings of the National Academy of Sciences* 101 (52): 18058–63.

Fortna, Andrew, Young Kim, Erik MacLaren, Kriste Marshall, Gretchen Hahn, Lynne Meltesen et al. 2004. Lineage-specific gene duplication and loss in human and great ape evolution. *PLOS Biology* 2 (7): e207.

Frisch, R. E. 1978. Population, food intake, and fertility: Historical evidence for a direct effect of nutrition on reproductive ability. *Science* 199:2–30.

———. 2002. *Female fertility and body fat connection.* Chicago: University of Chicago Press.

Galik, K., B. Senut, M. Pickford, D. Gommery, J. Treil, J. Kuperavage, and R. B. Eckhardt. 2004. External and internal morphology of the BAR 1002'00 *Orrorin tugenensis* femur. *Science* 305 (5689): 1450–53.

Ganguly, S., S. L. Coon, and D. C. Klein. 2002. Control of melatonin synthesis in the mammalian pineal gland: The critical role of serotonin acetylation. *Cell Tissue Research* 309 (1): 127–37.

Gani, Royhan M., and Nahid D. S. Gani. 2008. Tectonic hypotheses of human evolution. *Geotimes* 53 (1): 34–39.

Gaynor, Dave. *Animals under fire.* www.scarborough.org.za/ftp/animals%20 under%20fire-7.doc (accessed 2006).

Gilad, Yoav, Alicia Oshlack, Gordon K. Smyth, Terence P. Speed, and Kevin P. White. 2006. Expression profiling in primates reveals a rapid evolution of human transcription factors. *Nature* 440:242–45.

Gilbert, Scott F. 2000. *Developmental biology.* Sunderland, MA: Sinauer Associates. Also available online at NCBI. http://www.ncbi.nlm.nih. gov/books/bv.fcgi?rid=dbio.section.5122.

Glickman, Gena, John P. Hanifin, Mark D. Rollag, Jenny Wang, Howard Cooper, and George C. Brainard. 2003. Inferior retinal light exposure is more effective than superior retinal exposure in suppressing melatonin in humans. *Journal of Biological Rhythms* 18 (1): 71–79.

Goldman, Bruce. 2001. Mammalian photoperiodic system: Formal properties and neuroendocrine mechanisms of photoperiodic time measurement. *Journal of Biological Rhythms* 16 (4): 283–301.

Gommes, R. 2002. *Mountain climates.* Food and Agriculture Organization of the United Nations. http://www.fao.org/sd/2002/EN0701a_en.htm (accessed February 2006).

Goodall, Jane. 1968. The behavior of free-living chimpanzees in the Gombe stream reserve. *Animal Behavior Monographs* 1:165–311.

———. 1986. *The chimpanzee of Gombe: Patterns of behavior.* Boston: Belknap Press of Harvard University Press.

Goossens, B. M. A., M. Dekleva, S. M. Reader, E. H. M. Sterck, and J. J. Bolhuis. 2008. Gaze following in monkeys is modulated by observed facial expressions. *Animal Behaviour* 75 (5): 1673–81.

Goren-Inbar, Naama, Nira Alperson, Mordechai E. Kisleve, Orit Simchoni, Yoel Melamed, Adi Ben-Nun, and Ella Werker. 2004. Evidence of hominin control of fire at Gesher Benot Ya'aqov, Israel. *Science* 304:725–27.

Gorman, Michael R, and Theresa M. Lee. 2002. Hormones and biological rhythms. In *Behavioral endocrinology*, ed. J. B. Becker, S. M. Breedlove, David Crews, and Margaret M. McCarthy, 451–94. Cambridge, MA: MIT Press.

Gould, S. J. 1989. *Wonderful life: The burgess shale and the nature of history.* New York: W.W. Norton.

Govender, Navashni. 2003. Fire management in the Kruger National Park. *Aridlands Newsletter* 54. http://ag.arizona.edu/OALS/ALN/ALNHome. html.

Gowlett, J. A. J. 2006. The early settlement of northern Europe: Fire history in the context of climate change and the social brain. *Comptes Rendus Palévol* 5:299–310.

Gowlett, J .A. J., J. W. K. Harris, D. Waldton, and B. A. Wood. 1982. Early hominids and fire at Chesowanja, Kenya—Reply. *Nature* 296 (5960): 870.

Grafe, T. Ulmar, Stefanie Doebler, and K. Eduard Linsenmair. 2002. Frogs flee from the sound of fire. *Proceedings of the Royal Society Biological Sciences Series B* 269 (1495): 993–1003.

Green, Carla B., and Joseph C. Besharse. 2004. Retinal circadian clocks and control of retinal physiology. *Journal of Biological Rhythms* 19 (2): 91–102.

Greenspan, S. I., and S. G. E. Shanker. 2004. *The first idea: How symbols, language, and intelligence evolved from our primate ancestors to modern humans.* Cambridge, MA: De Capo Press.

Griffin, D. 1976. *The question of animal awareness: Evolutionary continuity of mental experience.* New York: The Rockefeller University Press.

———. 1984. *Animal thinking.* Cambridge, MA: Harvard University Press.

Grover, P. L., and F. L. Martin. 2002. The initiation of breast and prostate cancer. *Carcinogenesis* 23:1095–1102.

Groves, Colin, and J. Sabater Pi. 1985. From ape's nest to human fix-point. *Man* 20 (1): 22–47.

Grun, R., Chris Stringer, Frank McDermott, Roger Nathane, Naomi Porat, Steve Robertson et al. 2005. U-series and ESR analyses of bones and teeth relating to the human burials from Skhul. *Journal of Human Evolution* 49 (3): 316–34.

Gubbins, David, Adrian L. Jones, Christopher C. Finlay. 2006. Fall in earth's magnetic field is erratic. *Science* 312 (5775): 900–902.

Gullone, Eleonora. 2000. The development of normal fear: A century of research. *Clinical Psychology Review* 20 (4): 429–51.

Gumert, M. 2007. Payment for sex in a macaque mating market. *Animal Behaviour* 74:e1665–67.

Gunter, C., and R. Dhand. 2005. The chimpanzee genome. *Nature* 437.

Hahn, Matthew W., Jeffery P. Demuth, and Sang-Gook Han. 2007. Accelerated rate of gene gain and loss in primates. *Genetics* 177:1941–49.

Haile-Selassie, Yohannes. 2001. Late Miocene hominids from the Middle Awash, Ethiopia. *Nature* 412:178–81.

Hall, K. R. L. 1963. Observational learning in monkeys and apes. *British Journal of Psychology* 54:201–26.

———. 1968. *Social learning in monkeys.* In *Primates: Studies in adaptation and variability,* ed. P. C. Jay, 383–97. New York: Holt, Rhinehart, Winston.

Hare, Brian, and Michael Tomasello. 2005. Human-like social skills in dogs? *Trends in Cognitive Sciences* 9 (9): 439–44.

Harlow, Harry F. 1958. The nature of love. *American Psychologist* 13:673–85.

Harris, Michael. 1980. *Heating with wood.* Don Mills, Ontario: Citadel Press, General Publishing.

Hart, Donna L., and R. W. Susman. 2005. *Man the hunted: Primates, predators, and human evolution.* New York: Basic.

Harvey, Paul H., R. D. Martin, and T. H. Clutton-Brock. 1988. Life histories in comparative perspective. In *Primate societies,* ed. B. B. Smuts, D. L. Cheney, R. M. Seyfarth, R. W. Wrangham, and T. T. Struhsaker, 181–96. Chicago: University of Chicago Press.

Hashimoto, Satoko, Masako Kohsaka, Nobuyuki Morita, Noriko Fukuda, Sato Honma, and Kenichi Honma. 1996. Vitamin B12 enhances the phase-response of circadian melatonin rhythm to a single bright light exposure in humans. *Neuroscience Letters* 220 (2): 129–32.

Hatonen, T., A. Alila-Johansson, S. Mustanoja, and M. L. Laakso. 1999. Suppression of melatonin by 2000-lux light in humans with closed eyelids. *Biological Psychiatry* 46 (6): 827–31.

Hausfater, G., and J. B. Mead. 1982. Alternation of sleeping groves by yellow baboons (*Papio cynocephalus*) as a strategy for parasite avoidance. *Primates* 23:287–97.

Hawkes, K., J. F. O'Connell, N. G. Blurton Jones, H. Alvarez, and E. L. Charnov. 1998. Grandmothering, menopause, and the evolution of human life histories. *Proceedings of the National Academy of Sciences* 95 (3): 1336–39.

Heijmansa, B. T., E. W. Tobia, A. D. Sten, H. Putter, G. J. Baluw, E. S. Sussere et al. 2008. Persistent epigenetic differences associated with prenatal exposure to famine in humans. *PNAS* 105 (44): 17046–49.

Helfferich, Carla. 1996. Hardwood, softwood, fuelwood. *Alaska Science Forum, 1990.* http://www.gi.alaska.edu/ScienceForum/ASF13/1307.html (accessed 2005).

Hernandez-Aguilar, R. Adriana, Jim Moore, and Travis Rayne Pickering. 2007. Savanna chimpanzees use tools to harvest the underground storage organs of plants. *Proceedings of the National Academy of Sciences* 104 (49): 19210–13.

Heyes, Cecilia M. 1998. Theory of mind in nonhuman primates. *Behavioral and Brain Sciences* 21:101–48.

Hileman, Stanley M., Dominique D. Pierroz, and Jeffrey S. Flier. 2000. Leptin, nutrition, and reproduction: Timing is everything. *Journal of Clinical Endocrinological Metabolism* 85 (2): 804–7.

Hiraiwa-Hasegawa, M., R. W. Byrne, H. Takasaki, and J. M. Byrne. 1986. Aggression toward large carnivores by wild chimpanzees of Mahale Mountains National Park, Tanzania. *Folia Primatologica* 4 (1): 8–13.

Hirasaki, E., N. Ogihara, Y. Hamada, H. Kumakura, and M. Nakatsukasa. 2004. Do highly trained monkeys walk like humans? A kinematic study of bipedal locomotion in bipedally trained Japanese macaques. *Journal of Human Evolution* 46:739–50.

Hobson, J. Allan. 2005. Sleep is of the brain, by the brain, and for the brain. *Nature* 437:1254–56.

Hoberg, E. P., N. L. Alkire, A. de Queiroz, and A. Jones. 2001. Out of Africa: Origins of the *Taenia* tapeworms in humans. *Proceedings of the Royal Society B: Biological Sciences* 268:781–87.

Hohman, G., and B. Fruth. 2008. New records on prey capture and meat eating by bonobos at Lui Kotale, Salonga National Park, Democratic Republic of Congo. *Folia Primatologica* 79:103–10.

Honma, K., M. Kohsaka, N. Fukuda, N. Morita, and S. Honma. 1992. Effects of vitamin B12 on plasma melatonin rhythm in humans: Increased light sensitivity phase advances the circadian clock? *Experientia* 48 (8): 716–20.

Horack, John. *Combustion physics.* 1997. Marshall Space Flight Center. http://science.nasa.gov/ms11/combustion_why.htm.

House, J. I., and D. O. Hall. 2005. *Environmental determinants.* http://www.savannas.net/savenv.htm (accessed 2005).

Huffman, M. A. 1997. Current evidence for self-medication in primates: A multidisciplinary perspective. *Yearbook of Physical Anthropology* 40:171–200.

Huffman, Michael A., and Mohamedi Seifu Kalunde. 1993. Tool-assisted predation on a squirrel by a female chimpanzee in the Mahale Mountains, Tanzania. *Primates* 34 (1): 93–98.

Hutchinson, Sharon Elaine. n.d. *Nuer dilemmas: Coping with money, war, and the state.* Human Relations Area Files. http://www.yale.edu/hraf/ (accessed 2003).

Huttner, H., M. Beyer, and J. Bargon. 2007. Charcoal smoke causes bronchial anthracosis and COPD (article in German). *Med Klin (Munich)* 102 (1): 59–63.

Ikeda, M., M. Asai, T. Moriya, M. Sagara, S. Inoue, and S. Shibata. 1998. Methylcobalamin amplifies melatonin-induced circadian phase shifts by facilitation of melatonin synthesis in the rat pineal gland. *Brain Research* 795 (1–2): 98–104.

International Chimpanzee Chromosome Consortium. 2004. DNA sequence and comparative analysis of chimpanzee chromosome 22. *Nature* 429:382–88.

Irmak, M. K., T. Topal, and S. Oter. 2005. Melatonin seems to be a mediator that transfers the environmental stimuli to oocytes for inheritance of adaptive changes through epigenetic inheritance system. *Medical Hypotheses* 64 (6):1138–43.

Isaac, G. 1982. Early hominids and fire at Chesowanja, Kenya. *Nature* 296 (5860): 870.

Iżykowska, Ilona, Aleksandra Piotrowska, Marzena Podhorska-Okołów, Marek Cegielski, Maciej Zabel, and Piotr Dzięgiel. 2008. Ochronna rola melatoniny podczas działania promieniowania UV [The protective role of melatonin in the course of UV exposure]. *Postepy Hig Med Dosw* 62:23–27 (available online).

Jackendoff, Ray. 1999. Possible stages in the evolution of the language capacity. *Trends in Cognitive Sciences* 3 (7): 272–79.

James, S. R. 1989. Hominid use of fire in the Lower and Middle Pleistocene. *Current Anthropology* 30 (1): 1–26.

Kaessmann, Henrik, and Fabien Burki. 2004. Birth and adaptive evolution of a hominoid gene that supports high neurotransmitter flux. *Nature Genetics* 36:1061–63.

Kajobe, Roberta, and David W. B. Roubik. 2006. Honey-making bee colony abundance and predation by apes and humans in a Uganda forest reserve. *Biotropica* 38 (2): 210.

Kalloniatis, Michael, and Charles Luu. 2003. *Psychophysics of Vision, Part IX.* Salt Lake City: University of Utah, John Moran Eye Center. http://webvision.med.utah.edu/index.html (accessed 2003).

Kandel, E. R., J. H. Schwartz, and T. M. Jessell. 2000. *Principles of neural science.* 4th ed. New York: McGraw-Hill.

Kano, T. 1983. An ecological study of pygmy chimpanzees (*Pan paniscus*) of Yalosidi, Republic of Zaire. *International Journal of Primatology* 4:1–31.

Kano, T., and M. Mulavwa. 1984. Feeding ecology of pygmy chimpanzees (*Pan paniscus*) of Wamba. In *The pygmy chimpanzee: Evolutionary biology and behavior*, ed. R. L. Susman. New York: Plenum.

Keeley, Jon E., and Philip W. Rundel. 2005. Fire and the Miocene expansion of C4 grasslands. *Ecology Letters* 8:683–90.

Kimchi, Tali, Jennings Xu, and Catherine Dulac. 2007. A functional circuit underlying male sexual behaviour in the female mouse brain. *Nature* 448 (7157): 1009–14.

Klein, D. C. 2004. The 2004 Aschoff/Pittendrigh Lecture: Theory of the origin of the pineal gland—A tale of conflict and resolution. *Journal of Biological Rhythms* 19 (4): 264–79.

Klein, N., F. Fröhlich, and S. Krief. 2008. Geophagy: Soil consumption enhances the bioactivities of plants eaten by chimpanzees. *Naturwissenschaften* 95 (4): 325–31.

Klein, R. G. 1992. Archaeology and the evolution of human behavior. *Evolutionary Anthropology* 9 (1): 17–36.

Knott, Cheryl D. 2005. Energetic responses to food availability in the great apes: Implications for hominin evolution. In *Seasonality in primates: Studies of living and extinct human and nonhuman primates,* ed. D. K. Brockman and C. Van Schaik. Cambridge: Cambridge University Press.

Kobayashi, H., A. Kromminga, Thomas W. Dunlop, Birte Tychsen, Franziska Conrad, Naoto Suzuki et al. 2005. A role of melatonin in neuroectodermal-mesodermal interactions: The hair follicle synthesizes melatonin and expresses functional melatonin receptors. *FASEB Journal* 19 (12):1710–12.

Kolar, Jan, and Ivana Machackova. 2005. Melatonin in higher plants: Occurrence and possible functions. *Journal of Pineal Research* 39 (4): 333–41.

Kuller, R. 2002. The influence of light on circarhythms in humans. *Journal of Physiological Anthropology and Applied Human Science* 21 (2): 87–91.

Kummer, H. 1968. *Social organization of hamadryas baboons.* Vol. 6 of *Bibl. Primat.* Basel: S. Karger.

———. 1971. *Primate societies: Group techniques of ecological adaptation.* Chicago: Aldine-Atheron.

Kunimatsu, Yutaka, Masato Nakatsukasa, Yoshihiro Sawada, Sakai Testsuya, Msayuki Hyodo, Hironobu Hyodo et al. 2007. A new Late Miocene great ape from Kenya and its implications for the origins of African great apes and humans. *PNAS* 104 (49): 19220–25.

Kunzig, Robert. 2001. The physics of fire: Infernal combustion: After eons of sputtering research, the science of fire goes into orbit. *Discover* 22 (1): 35.

Kuroda, S. 1984. Interaction over food among pygmy chimpanzees. In *The pygmy chimpanzee: Evolutionary biology and behavior,* ed. R. L. Susman. New York: Plenum.

Lam, Raymond W. 1994. Seasonal affective disorder. *Current Opinion in Psychiatry* 7 (1): 9–13.

Leakey, Meave G., Fred Spoor, Frank H. Brown, Patrick N. Gathogo, Christopher Kiarie, Louise N. Leakey, and Ian McDougall. 2001. New hominin genus from eastern Africa shows diverse middle Pliocene lineages. *Nature* 410 (6827): 433–40.

Ledford, H. 2008. Human genes are multitaskers: Up to 94 percent of human genes can generate different products. *Nature News.* http://www.nature.com/news/2008/081102/full/news.2008.1199.html.

Lee, Sue-Hyun, Jun-Hyeok Choi, Nuribalhae Lee, Hye-Ryeon Lee, Jae-Ick Kim, Nam-Kyung Yu et al. 2008. Synaptic protein degradation underlies destabilization of retrieved fear memory. *Science* 319:1253–56.

Leonard, William R. 2002. Food for thought: Dietary change was a driving force in human evolution. *Scientific American* 287 (6): 106–16.

Leonard, William R., and Marcia L. Robertson. 1992. Nutritional requirements and human evolution: A bioenergetics model. *American Journal of Human Biology* 4 (2): 179–95.

———. 1994. Evolutionary perspectives on human nutrition: The influence of brain and body size on diet and metabolism. *American Journal of Human Biology* 6 (1): 77–88.

Lieberman, P., D. H. Klatt, and W. H. Wilson. 1969. Vocal tract limitations on the vowel repertoires of rhesus monkey and other nonhuman primates. *Science* 164:1185–87.

Liebmann, P. M., A. Wolfler, P. Felsner, D. Hofer, and K. Schauenstein. 1997. Melatonin and the immune system. *Int. Arch. Allergy Immunol.* 112 (3): 203–11.

Liman, E. R., and H. Innan. 2003. Relaxed selective pressure on an essential component of pheromone transduction in primate evolution. *PNAS* 100 (6): 3328–32.

Lincoln, G. A., and M. Richardson. 1998. Photo-neuroendocrine control of seasonal cycles in body weight, pelage growth, and reproduction: Lessons from the HPD sheep model. *Comparative Biochemistry and Physiology—Part C: Pharmacology, Toxicology and Endocrinology* 119 (3): 283–94.

Linford, N. T. 2001. Geophysical evidence for fires in antiquity: Preliminary results from an experimental study. *Archaeological Prospection, Exeter* 8 (4): 211–25.

Lockley, S. W., G. C. Brainard, and C. A. Czeisler. 2003. High sensitivity of the human circadian melatonin rhythm to resetting by short wavelength light. *Journal of Clinical Endocrinology and Metabolism* 88 (9): 4502–5.

Lonergan, N. 2008. The hominin fossil record: Taxa, grades, and clades. *Journal of Anatomy* 212 (4): 354–76.

Lucas, P. W., P. J. Constantino, and B. Wood. 2008. Inferences regarding the diet of extinct hominins: Structural and functional trends in dental and mandibular morphology within the hominin clade. *Journal of Anatomy* 212 (4): 486–500.

Luvone, P. Michael, Gianluca Tosini, Nikita Pozdeyev, Rashidul Haque, David C. Klein, and Shyam S. Chaurasia. 2005. Circadian clocks, clock networks, arylalkylamine N-acetyltransferase, and melatonin in the retina. *Progress i n Retinal and Eye Research* 24 (4): 433–56.

Maclatchy, Laura 2004. The oldest ape. *Evolutionary Anthropology* 13 (3): 90–103.

Marshall-Pescini, Sarah, and Andrew Whiten. 2008. Chimpanzees (*Pan troglodytes*) and the question of cumulative culture: An experimental approach. *Animal Cognition* 11 (3): 449–56.

Martinez, M., J.-L. Rosa, P. Arsuaga, R. Jarabo, C. Quam, A. Lorenzo et al. 2004. Auditory capacities in Middle Pleistocene humans from the Sierra de Atapuerca in Spain. *Proceedings of the National Academy of Sciences* 101 (27): 9976–81.

Maslin, Mark A., and Beth Christensen. 2007. African paleoclimate and human evolution. *Journal of Human Evolution* 53 (5): 443–64.

Mazur, Witold. 1998. Phytoestrogen content in foods. *Bailliere's Clinical Endocrinology and Metabolism* 12 (4): 729–42.

McDougall, Ian, Francis H. Brown, and John G. Fleagle. 2005. Stratigraphic placement and age of modern humans from Kibish, Ethiopia. *Nature* 433:733–36.

McGraw, W. Scott, Catherine Cooke, and Susanne Schultz. 2006. Primate remains from African crowned eagle (*Stephanoaetus coronatus*) nests in Ivory Coast's Tai Forest: Implications for primate predation and early hominid taphonomy in South Africa. *American Journal of Physical Anthropology* 131(2): 151–65.

McGrew, C. W. 1989. Comment on hominid use of fire in the Lower and Middle Pleistocene by S. R. James. *Current Anthropology* 30(1):16–17.

———. 1992. *Chimpanzee natural culture: Implications for human evolution.* Cambridge: Cambridge University Press.

———. 1994. Tools compared: The material of culture. In *Chimpanzee cultures*, ed. R. W. Wrangham, C. W. McGrew, F. B. M. de Waal, and P. Heltne. Cambridge, MA: Harvard University Press.

———. 1998. Culture in nonhuman primates? *Annual Review of Anthropology* 27:301–28.

McGrew, W. C., J. D. Pruetz, and S. J. Fulton. 2005. Chimpanzees use tools to harvest social insects at Fongoli, Senegal. *Folia Primatologica* 76 (4): 222–26.

references

Méndez-Ferrer, Simón, Daniel Lucas, Micela Battista, and Paul S. Frenette. 2008. Haematopoietic stem cell release is régulated by circadian oscillations. *Nature* 452:442–47.

Mercader, Julio, Melissa Panger, and Christophe Boesch. 2002. Excavation of a chimpanzee stone tool site in the African rainforest. *Science* 296 (5572): 1452–55.

Milton, Katharine. 1993. Diet and primate evolution. *Scientific American* 269 (2): 86–93.

———. 1999. A hypothesis to explain the role of meat-eating in human evolution. *Evolutionary Anthropology* 8 (1): 11–21.

———. 2003. The critical role played by animal source foods in human (*Homo*) evolution 1, 2. *Journal of Nutrition* 133 (11) Suppl. no. 2: 3886S-92S.

Moore-Ede, Martin C., and L. Margaret Moline. 1985. Circadian rhythms and photoperiodism. In *Photoperiodism, melatonin, and the pineal.* CIBA Foundation Symposium 117. Pitman: London. Also online at http://www.ncbi.nlm.nih.gov/pubmed/3836815 (accessed 2006).

Moyà-Solà, Salvador, Meike Köhler, David M. Alba, and Isaac Casanovas-Vilar. 2004. *Pierolapithecus catalaunicus*, a new Middle Miocene great ape from Spain. *Science* 306 (5700): 1339–45.

Münch, Mirjam, Szymon Kobialka, Roland Steiner, Peter Oelhafen, Anna Wirz-Justice, and Christian Cajochen. 2006. Wavelength-dependent effects of evening light exposure on sleep architecture and sleep EEG power density in men. *American Journal of Physiology—Regulatory, Integrative, and Comparative Physiology* 290:R1421–28.

Munckenbeck-Fragaszy, D., and G. Mitchell. 1974. Infant socialization in primates. *Journal of Human Evolution.* 3:563–75.

Murch, Susan J., Skye S. B. Campbell, and Praveen K. Saxena. 2001. The role of serotonin and melatonin in plant morphogenesis: Regulation of auxin-induced root organogenesis in in vitro–cultured explants of St. John's Wort (Hypericum Perforatum L.). *In Vitro Cellular and Development Biology—Plant* 37(6): 786–93.

Mustonen, A. M., P. Nieminen, and H. Hyvarinen. 2001. Preliminary evidence that pharmacologic melatonin treatment decreases rat ghrelin levels. *Endocrine* 16 (1): 43–46.

Nakamichi, M., N. Itoigawa, S. Imakawa, and S. Machida. 2005. Dominance relations among adult females in a free-ranging group of Japanese monkeys at Katsuyama. *American Journal of Primatology* 37 (3): 241–51.

Naskrecki, Piotr. 2005. *The smaller majority*. Cambridge, MA: Harvard University Press.

Nguyen, Dzung H., Nancy Hurtado-Ziola, Pascal Gagneux, and Ajit Varki. 2006. Loss of Siglec expression on T lymphocytes during human evolution. *Proceedings of the National Academy of Sciences* 103 (20): 7765–70.

Nishida, T. 1979. *The social structure of the chimpanzee of the Mahale Mountains*. In *The great apes*, ed. D. A. Hamburg and E. R. McCown. Menlo Park, CA: Benjamin/Cummings.

Nishimura, Takeshi, Akichika Mikami, Juri Suzuki, and Tetsuro Matsuzawa. 2003. Descent of the larynx in chimpanzee infants. *PNAS* 100 (12): 6930–33.

Oakley, Kenneth Page. 1955. Fire as paleolithic tool and weapon. *Prehistoric Society Proceedings*, n.s., 21:36–48.

———. 1961. On man's use of fire, with comments on tool-making and hunting. In *Social life of early man*, ed. S. L. Washburn. New York: Viking Fund Publications in Anthropology.

Oldham, Michael C., Steve Horvath, and Daniel H. Geschwind. 2006. Conservation and evolution of gene coexpression networks in human and chimpanzee brains. *Biological Sciences—Neuroscience* 103 (47): 17973–78.

Paabo, Svante. 2003. The mosaic that is our genome. *Nature* 421:409–12.

Palmer, D. P., P. Pettit, and P. G. Bahn. 2005. *Unearthing the past: The great archaeological discoveries that have changed history*. Philadelphia: Running Press.

Pang, S. F., N. T. Linford, E. A. Ayre, C. S Pang, P. P. Lee, R. K. Xu et al. 1998. Neuroendocrinology of melatonin in reproduction: Recent developments. *Journal of Chemical Neuroanatomy* 14 (3–4): 157–66.

Parent, Anne-Simone, Grete Teilmann, Anders Juul, Niels E. Skakkebaek, Jorma Toppari, and Jean-Pierre Bourguignon. 2003. The timing of normal puberty and the age limits of sexual precocity: Variations around the world, secular trends, and changes after migration. *Endocrine Reviews* 24 (5): 668–93.

Patterson, N., Daniel J. Richter, Sante Gnerre, Eric S. Lander, and David Reich. 2006. Genetic evidence for complex speciation of humans and chimpanzees. *Nature* 441:1103–8.

Perlès, C. 1977. *Préhistoire du feu*. Paris: Masson.

Perry, George H., Brian C. Verreli, and Anne C. Stone. 2005. Comparative analyses reveal a complex history of molecular evolution for human *MYH16*. *Molecular Biology and Evolution* 22 (3): 379–82.

Peters, Charles R., and John C. Vogel. 2005. Africa's wild C4 plant foods and possible early hominid diets. *Journal of Human Evolution* 48 (3): 219–36.

Petraglia, Michael D. 2002. The heated and the broken: Thermally altered stone, human behavior, and archaeological site formation. *North American Archaeologist* 23 (3): 241–69.

Pfeiffer, John. 1971. When *Homo erectus* tamed fire, he tamed himself. In *Readings in general anthropology*, ed. L. D. Holmes. New York: Ronald Press.

Phelps, Elizabeth A., Mauricio R. Delgado, Katherine I. Nearing, and Joseph E. LeDoux. 2004. Extinction learning in humans: Role of the amygdala and vmPFC. *Neuron* 43 (6): 897–905.

Phillipson, David W. 2005. *African archaeology.* 3rd ed. Cambridge: Cambridge University Press.

Piperno, D. R., and H. D. Sues. 2005. Paleontology: Dinosaurs dined on grass. *Science* 310 (5751): 1126–28.

Plachetzki, David C., Jeanne M. Serb, and Todd H. Oakley. 2005. New insights into the evolutionary history of photoreceptor cells. *Trends in Ecology and Evolution* 20 (9): 465–67.

Platek, S. M., G. G. Gallup Jr., and B. D. Fryer. 2002. The fireside hypothesis: Was there differential selection to tolerate air pollution during human evolution? *Medical Hypotheses* 58 (1): 1–5.

Pollard, Katherine S., Sofie R. Salama, Nelle Lambert, Marie-Alexandra Lambot, Sandra Coppens, Jakob S. Pedersen et al. 2006. An RNA gene expressed during cortical development evolved rapidly in humans. *Nature* 443:167–72.

Popesco, Magdalena, Erik J. MacLaren, Janet Hopkins, Laura Dumas, Michael Cox, Lynne Meltesen et al. 2006. Human lineage–specific amplification, selection, and neuronal expression of duf1220 domains. *Science* 313 (5791): 1304–7.

Povinelli, Daniel J., Laura A. Theall, James E. Reaux, and Sarah Dunphy-Lelii. 2003. Chimpanzees spontaneously alter the location of their gestures to match the attentional orientation of others. *Animal Behaviour* 66 (1): 71–79.

Prathibha, S., B. Nambisan, and S. Leelamma. 1995. Enzyme inhibitors in tuber crops and their thermal stability. *Plant Foods for Human Nutrition* 48 (3): 247–57.

Preuss, Todd M., Mario Cáceres, Michael C. Oldham, and Daniel H. Geschwind. 2004. Human brain evolution: Insights from microarrays. *Nature Reviews Genetics* 5:850–60.

Price, David M. 2003. *Thermoluminescence dating*. School of Geosciences. http://www.ansto.com.au/ainse/general2003/quaternary_wkshp_2003_01.pdf.

Pruetz, J. D. 2001. Use of caves by Savanna chimpanzees (*Pan troglodytes verus*) in the Tomboronkoto Region of southeastern Senegal. *Pan Africa News* 8 (2): 26–28.

Pruetz, Jill D., and Paco Bertolani. 2007. Savanna chimpanzees, *Pan troglodytes verus*, hunt with tools. *Current Biology* 17 (5, 6): 412–17.

Raaum, R. L., K. N. Sterner, Colleen M. Noviello, Caro-Beth Stewart, and Todd R. Disotell. 2005. Catarrhine primate divergence dates estimated from complete mitochondrial genomes: Concordance with fossil and nuclear DNA evidence. *Journal of Human Evolution* 48 (3): 237–57.

Ragir, Sonia, Martin Rosenberg, and Phil Tierno. 2000. Gut morphology and the avoidance of carrion among chimpanzees, baboons, and early hominids. *Journal of Anthropological Research* 56:477–512.

Raloff, J. 1998. Does light have a dark side? Nighttime illumination might elevate cancer risk. *Science News Online* 154 (16): 1–7.

Ramachandran, V. S. *Mirror neurons*. 2005. www.edge.org/3rd_culture/ramachandran/ramchandran_index.html.

Rantala, Markus J. 1999. Controversies in parasitology: Human nakedness: Adaptation against ectoparasites? *International Journal for Parasitology* 29:1987–89.

Rawashdeh, Oliver, Nancy Hernandez de Borsetti, Gregg Roman, and Gregory M. Cahill. 2007. Melatonin suppresses nighttime memory formation in Zebrafish. *Science* 318 (5853): 1144–46.

Rea, Mark. 2002. Light: Much more than vision. In *EPRI/LRO 5th International Lighting Research Symposium*. Palo Alto, CA: Electric Power Research Institute Lighting Research Office.

Read, Tamar R., and Sean M. Bellairs. 2003. Smoke affects the germination of native grasses of New South Wales. *Australian Journal of Botany* 47 (4): 563–76.

Reed, David K., Jessica E. Light, Julie M. Allen, and Jeremy J. Kirchman. 2007. Pair of lice lost or parasites regained: The evolutionary history of anthropoid primate lice. *BMC Biology* 5:7.

Reed, K. E., and J. L. Fish. 2005. Tropical and temperate seasonal influences on human evolution. In *Seasonality in primates: Studies of living and extinct human and nonhuman primates*, ed. D. K. Brockman and C. van Schaik. Cambridge: Cambridge University Press.

Reppert, Steven M., Marilyn J. Duncan, and Bruce D. Goldman. 1985. Photic influences on the developing mammal. *Symposium 117 on Melatonin and the Pineal*, 116-28. CIBA Foundation. London: Pitman.

Riesenfeld, Alphonse. 1966. The effects of experimental bipedalism and upright posture in the rat and their significance for the study of human evolution. *Acta anatomica* 65:449–521.

Rilling, J. K., M. F. Glasser, T. M. Preuss, X. Ma, T. Zhao, X. Hu et al. 2008. The evolution of the arcuate fasciculus revealed with comparative DTI. *Nature Neuroscience* 11:426–28.

Rizzolatti, Giacomo, and Michael A. Arbib. 1998. Language within our grasp. *Trends in Neurosciences* 21 (5): 188–94.

Rockman, Matthew V., Matthew W. Hahn, Nicole Soranzo, Fritz Zimprich, and David B. Goldstein. 2005. Ancient and recent positive selection transformed opioid cis-regulation in humans. *PLOS Biology* 3 (2): 1–12.

Roennenberg, Till. 2004. The decline in human seasonality. *Journal of Biological Rhythms* 19 (3): 193–95.

Roenneberg, Till, Tim Kuehnle, Peter P. Pramstaller, Jan Ricken, Miriam Havel, Angelika Guth, and Martha Merrow. 2004. A marker for the end of adolescence. *Current Biology* 14 (24): R1038–39.

Roennenberg, Till, Serge Daan, and Martha Merrow. 2003. Basics: The art of entrainment. *Journal of Biological Rhythms* 18 (3): 183–94.

Rogers, A. R., D. Iltis, and Stephen Wooding. 2004. Genetic variation at the *MC1R* locus and the time since loss of human body hair. *Current Anthropology* 45 (1): 105–8.

Rolland, Nicolas. 2004. Was the emergence of home bases and domestic fire a punctuated event? A review of the Middle Pleistocene record in Eurasia. *Asian Perspectives* 43 (2): 248–81.

Ronen, Avraham. 1998. Domestic fire as evidence for language. In *Neandertals and modern humans in western Asia*, ed. T. Akazawa, K. Aoiki, and O. Bar-Yosef. New York: Plenum.

Rose, L. M. 1997. *Cebus* meets *Pan*. *International Journal of Primatology* 18 (5): 727–65.

Rose, M. D. 1991. The process of bipedalization in hominids. In *Origine(s) de la Bipedie Chez les Hominides*, ed. B. Senut and Y. Coppens. Paris: CNRS.

Ruger, M., M. C. Gordijn, D. G. Beersma, B. de Vries, and S. Daan. 2005. Nasal versus temporal illumination of the human retina: Effects on core body temperature, melatonin, and circadian phase. *Journal of Biological Rhythms* 20 (1): 60–70.

Sadoski, M., A. Paivio, and E. T. Goetz. 1991. A critique of schema theory in reading and a dual coding alternative. *Reading Research Quarterly* 26:463–84.

Sancar, A. 2000. Cryptochrome: The second photoactive pigment in the eye and role in circadian photoreception. *Annual Reviews Biochemistry* 69:31–67.

———. 2004. Regulation of the mammalian circadian clock by cryptochrome. *Journal of Biological Chemistry* 279 (33): 34079–82.

Sanz, Crickette, Dave Morgan, and Steve Gulick. 2004. New Insights into chimpanzees, tools, and termites from the Congo Basin. *American Naturalist* 164:567–81.

Sarafis, Vassilios, and Maciej Henneberg. 1999. Taeniid parasite evidence for eating meat as a natural part of the human diet. *Perspectives in Human Biology* 4 (1): 47–49.

Schmidt, Timothy R., Allon Goldberg, Derek E. Wildman, Timothy R. Schmidt, Maik Hüttemann, Morris Goodman et al. 2003. Adaptive evolution of cytochrome c oxidase subunit VIII in anthropoid primates. *Proceedings of the National Academy of Sciences* 100:5873–78.

Schultz, Adolph. 1969. *The life of primates.* New York: Universe Books.

Science in Africa. 2004. http://www.scienceinafrica.co.za/2004/april/braai.htm.

Scott, A. C. 2000a. Introduction to fire and the palaeoenvironment. *Palaeogeography, Palaeoclimatology, Palaeoecology* 164 (1–4): 7–11.

———. 2000b. The Pre-Quaternary history of fire. *Palaeogeography, Palaeoclimatology, Palaeoecology* 164 (1–4): 297–345.

Scott, Robert S., Peter S. Ungar, Torbjorn S. Bergstrom, Christopher A. Brown, Frederick E. Grine, Mark F. Teaford, and Alan Walker. 2005. Dental microwear texture analysis shows within-species diet variability in fossil hominins. *Nature* 436:693–95.

Sellers, William I., Gemma M. Cain, Weijie Wang, and Robin H. Crompton. 2005. Stride lengths, speed, and energy costs in walking of *Australopithecus afarensis*: Using evolutionary robotics to predict locomotion of early human ancestors. *J. R. Soc. Interface* 2:431–41.

Sellers, W. I., D. A. Dennis, and R. H. Crompton. 2003. Predicting the metabolic energy costs of bipedalism using evolutionary robotics. *Journal of Experimental Biology* 206:1127–36.

Senut, B., M. Pickford, D. Gommery, P. Mein, K. Cheboi, and Y. Coppens. 2001. First hominid from the Miocene (Lukeino Formation, Kenya). *Comptes Rendus de l'Academie des Sciences IIA* 332:137–44.

Shapiro, Harry L. 1975. *Peking man*. New York: Simon and Schuster.

Shapiro, L. J., and W. L. Jungers. 1988. Back muscle function during bipedal walking in chimpanzee and gibbon: Implications for the evolution of human locomotion. *American Journal of Physical Anthropology* 77 (2): 201–12.

Shen, G., T. Ku, Hai Cheng, Zhenxin Yuan, R. Lawrence Edwards, and Qian Wang. 2001. High-precision U-series dating of Locality 1 at Zhoukoudian, China. *Journal of Human Evolution* 41(6): 679–88.

Sherry, D. S. 2002. Reproductive seasonality in wild chimpanzees: A new method of analysis from Kibale, Uganda. *American Journal of Physical Anthropology* 34:140–41.

Shiu, S. Y. W., A. M. S. Poon, G. M. Brown, and S. F. Pang. 1998. Melatonin receptors in reproductive tissues: Evidence for the multiple sites of melatonin action. Paper presented at INABIS 1998—5th Internet World Congress on Biomedical Sciences, McMaster University, Canada.

Shostak, Marjorie. 1981. *Nisa: The life and words of a !Kung woman*. Cambridge, MA: Harvard University Press.

Shubin, N. H., E. B. Daeschler, and F. A. Jenkins. 2006. The pectoral fin of *Tiktaalik roseae* and the origin of the tetrapod limb. *Nature* 440:764–71.

———. 1998. Culture in nonhuman primates? *Annual Review of Anthropology* 27:301–28.

Siegel, J. M. 2005. Clues to the functions of mammalian sleep. *Nature* 437:1264–71.

Sisk, Cheryl L., and Douglas L. Foster. 2004. The neural basis of puberty and adolescence. *Nature Neuroscience* 7:1040–47.

Skinner, A., J. Lloyd, C. Brain, and F. Thackeray. 2004. Electron spin resonance and the first use of fire. Paper presented at the meeting of the Paleoanthropology Society, Montreal Canada.

Slominski, A., T. W. Fischer et al. 2005. On the role of melatonin in skin physiology and pathology. *Endocrine* 27 (2): 137–48.

Sponheimer, Matt, Julia Lee-Thorp, Darryl de Ruiter, Daryl Codron, Jacqui Codron et al. 2005. Hominins, sedges, and termites: New carbon isotope data from the Sterkfontein Valley and Kruger National Park. *Journal of Human Evolution* 48 (3): 301–12.

Stahl, A. B. 1984. Hominid diet before fire. *Current Anthropology* 25:151–68.

Stanford, Craig. 2003. *Upright: The evolutionary key to becoming human.* New York: Houghton-Mifflin.

———. 2006. *The predatory behavior and ecology of wild chimpanzees.* http://www-ref.usc.edu/~stanford/chimphunt.html.

Stanford, C. B., C. Gambaneza, J. B. Nkurunungi, and M. L. Goldsmith. 2000. Chimpanzees in Bwindi-Impenetrable National Park, Uganda, use different tools to obtain different types of honey. *Primates* 41:337–41.

Stedman, H. H., Benjamin W. Kozyak, Anthony Nelson, Danielle M. Thesier, Leonard T. Su, and David W. Low. 2004. Myosin gene mutation correlates with anatomical changes in the human lineage. *Nature* 428:415–18.

Stehle, J. H., C. Von Gall, and H. W. Korf. 2003. Melatonin: A clock-output, a clock-input. *Journal of Neuroendocrinology* 15:383–89.

Steiger, Axel. 2004. Eating and sleeping—their relationship to ghrelin and leptin. *American Journal of Physiology—Regulatory, Integrative, and Comparative Physiology* 287:R1031–32.

Steudel-Numbers, Karen L. 2003. The energetic cost of locomotion: Humans and primates compared to generalized endotherms. *Journal of Human Evolution* 44:255–62.

Stewart, Caro-Beth, and R. Todd Disotell. 1998. Primate evolution—in and out of Africa. *Current Biology* 8 (16): R582–88.

Stewart, O. C. 1958. Fire as the first great force employed by man. In *Man's role in changing the face of the earth*, ed. J. W. L. Thomas. Chicago: University of Chicago Press.

Straus, L. G. 1989. On early hominid use of fire. *Current Anthropology* 30 (4): 488–91.

Struhsaker, Thomas T. 1967. *Behavior of vervet monkeys (Cercopithecus aethiops).* Berkeley: University of California Press.

———. 1975. *The red colobus monkey.* Chicago: University of Chicago Press.

Struhsaker, T. T., D. O. Cooney, and K. S. Siex. 1997. Charcoal consumption by Zanzibar red colobus monkeys: Its function and its ecological and demographic consequences. *International Journal of Primatology* 18 (1): 61–72.

Strum, Shirley C. 1981. Processes and products of change: Baboon predatory behavior at Gilgil, Kenya. *Omnivorous primates: Gathering and hunting in human evolution*, ed. R. S. O. Harding and G. Teleki. New York: Columbia University Press.

———. 1987. *Almost human: A journey into the world of baboons*. New York: Random House.

———. 1998. *Baboon tales*. United States: Bullfrog Films.

Strum, S. C., and William Mitchell. 1987. Baboon models and muddles. In *The evolution of human behavior: Primate models*, ed. W. Kinzey. Albany: State University of New York Press.

Surbecka, M., and G. Hohmann. 2008. Primate hunting by bonobos at Lui Kotale, Salonga National Park. *Current Biology* 18 (19): R906–7.

Susman, Randall L. 1987. Chimpanzees: Pygmy chimpanzees and common chimpanzees: Models for the behavioral ecology of the earliest hominids. *The evolution of human behavior: Primate models*. W. Kinzey ed. Albany: State University of New York Press.

———. 1991. Species attribution of the Swartkrans thumb metacarpals: Reply to Drs. Trinkaus and Long. *American Journal of Physical Anthropology* 86 (4): 549–52.

Suwa, G., R. T. Kono, S. Katoh, B. Asfaw, and Y. Beyene. 2007. A new species of great ape from the late Miocene epoch in Ethiopia. *Nature* 448:921–24.

Svensen, Henrik, and Dag Kristian Dysthe. 2003. Subsurface combustion in Mali: Refutation of the active volcanism hypothesis in West Africa. *Geology* 31 (7): 581–84.

Taglialatela, Jared P., Jamie L. Russell, Jennifer A. Schaeffer, and William D. Hopkins. 2008. Communicative signaling activates "Broca's" homolog in chimpanzees. *Current Biology* 18:343–48.

Talbot, Lee Merriam. n.d. *Ecology of western Masailand, east Africa, File Maasai FL12*. Yale University. http://www.yale.edu/hraf/ (accessed 2003).

Tamura, Hiroshi, Yasuhiko Nakamura, Akio Narimatsu, Yoshiaki Yamagata, Akihisa Takasaki et al. 2008. Melatonin treatment in peri- and postmenopausal women elevates serum high-density lipoprotein cholesterol levels without influencing total cholesterol levels. *Journal of Pineal Research Online*. doi:10.1111/j.1600–079X.2008.00561.x.

Tavare, Simon, Charles R. Marshall, Oliver Will, Christophe Soligo, and Robert D. Martin. 2002. Using the fossil record to estimate the age of the last common ancestor of extant primates. *Nature* 416:726–29.

Teaford, Mark, and Peter Ungar. 2000. Diet and evolution of earliest human ancestors. *Proceedings of the National Academy of Sciences* 97:13506–11.

Teleki, Geza. 1981. The omnivorous diet and eclectic feeding habits of chimpanzees in Gombe National Park, Tanzania. In *Omnivorous primates: Gathering and hunting in human evolution*, ed. R. S. O. Harding and G. Teleki. New York: Columbia University Press.

Thieme, H. 1997. Lower Palaeolithic hunting spears from Germany. *Nature* 385: 807–10.

Thomas, Elizabeth Marshall. 1959. *The harmless people*. New York: Knopf.

Tillman, David A. 1978. *Wood as an energy resource*. New York: Academic Press.

Tomasello, Michael. 1994. The question of chimpanzee culture. In *Chimpanzee cultures*, ed. R. W. Wrangham, C. W. McGrew, F. B. M. de Waal, and P. Heltne. Cambridge, MA: Harvard University Press.

Tomasello, Michael, Josep Call, and Brian Hare. 2003. Chimpanzees understand psychological states—the question is which ones and to what extent. *Trends in Cognitive Sciences* 7 (4): 153–56.

Torres-Farfan, C., V. Rocco, C. Monsó, F. J. Valenzuela, C. Campino, A. Germain et al. 2006. Maternal melatonin effects on clock gene expression in a nonhuman primate fetus. *Endocrinology* 147 (10): 4618–26.

Tsukahara, Takahiro. 2005. Lions eat chimpanzees: The first evidence of predation by lions on wild chimpanzees. *American Journal of Primatology* 29 (1): 1–11.

Underdown, S. 2006. How the word "hominid" evolved to include hominin. *Nature* 444:680.

USGS. 2006a. *5.0. A standard national vegetation classification system*. United States Government. http://biology.usgs.gov/npsveg/classification/sect5.html (accessed April 2006).

———. 2006b. *Effects of fire in the Northern Great Plains*. United States Government. http://www.npwrc.usgs.gov/resource/habitat/fire (accessed 2006).

Van Gelder, R. N. 2003. Making sense of nonvisual ocular photoreception. *Trends in Neuroscience* 26 (9): 458–61.

Van Hoof, Jan A. R. A. M. 1994. Understanding chimpanzee understanding. In *Chimpanee cultures*, ed. R. W. Wrangham, C. W. McGrew, F. B. M. de Waal, and P. Heltne. Cambridge, MA: Harvard University Press.

Van Schaik, Carel, Richard Madden, and Jorg U. Ganzhorn. 2005. Seasonality and primate communities. In *Seasonality in primates: Studies of living and extinct human and nonhuman primates*, ed. D. K. Brockman and C. Van Schaik. Cambridge: Cambridge University Press.

Vandewalle, G., S. Gais, M. Schabus, E. Balteau, J. Carrier, A. Darsaud et al. 2007. Wavelength-dependent modulation of brain responses to a working memory task by daytime light exposure. *Cerebral Cortex* 17 (12): 2788–95.

Vanecek, Jiri. 1999. Inhibitory effect of melatonin on GnRh-induced LH release. *Reviews of Reproduction* 4:67–72.

Vargha-Khadem, Faraneh, David G. Gadian, Andrew Copp, and Mortimer Mishkin. 2005. *FOXP2* and the neuroanatomy of speech and language. *Nature Reviews Neuroscience* 6:131–38.

Verhaegen, M., and P. F. Puech. 2000. Hominid lifestyle and diet reconsidered: Paleoenvironmental and comparative data. *Human Evolution* 15 (3–4): 175–86.

Videan, Elaine N., and W. C. McGrew. 2002. Bipedality in chimpanzee (*Pan troglodytes*) and bonobo (*Pan paniscus*): Testing hypotheses on the evolution of bipedalism. *American Journal of Physical Anthropology* 118 (2): 184–90.

Vogel, Erin R., Janneke T. Van Woerden, Peter W. Lucas, Sri S. Utami Atmoko, Carel P. van Schaik, and Nathaniel J. Dominy. 2008. Functional ecology and evolution of hominoid molar enamel thickness: *Pan troglodytes schweinfurthii* and *Pongo pygmaeus wurmbii*. *Journal of Human Evolution* 55 (1): 60–74.

Vogel, Gretchen. 2002. Can chimps ape ancient hominid toolmakers? *Science* 296 (5572): 1380.

Vogel, Michael. 2003. *Heating with wood: Principles of combustion*. Montana State University Extension Service. http://www.montana.edu/wwwpb/pubs/ mt8405.html.

von Gall, Charlotte, Martine L. Garabette, Christian A. Kell, Sascha Frenzel, Faramarz Dehghani, Petra-Maria Schumm-Draeger et al. 2002. Rhythmic gene expression in pituitary depends on heterologous sensitization by the neurohormone melatonin. *Nature Neuroscience* 5:234–38.

Wakayama, E. J., J. W. Dillwith, R. W. Howard, and G. J. Blomquist. 1984. Vitamin B12 levels in selected insects. *Insect Biochemistry* 14 (2): 175–79.

Wang, Weijie, Robin H. Crompton, Tanya S. Carey, Yu Lia Gunther, Russell Savage, and William Sellers. 2004. Comparison of inverse-dynamics musculo-skeletal models of AL 288-1 *Australopithecus afarensis* and KNM-WT 15000 *Homo ergaster* to modern humans, with implications for the evolution of bipedalism. *Journal of Human Evolution* 47:453e478.

Waterland, R. A., and R. L. Jirtle. 2003. Transposable elements: Targets for early nutritional effects on epigenetic gene regulation. *Molecular and Cellular Biology* 23 (15): 5293–5300.

Wehr, T. A. 2001. Photoperiodism in humans and other primates: Evidence and implications. *Journal of Biological Rhythms* 16 (4): 348–64.

Weiss, Daniel J., Jason D. Wark, and David A. Rosenbaum. 2007. Monkey see, monkey plan, monkey do: The end-state comfort effect in cotton-top tamarins (*Saguinus oedipus*). *Psychological Science* 18 (12): 1063–68.

Welker, Glenn. 2005. *Apache—fire.* http://www.bedtime-story.com/bedtime-story/apache-fire.htm.

Wertsa, S. P., and A. H. Jahren. 2007. Estimation of temperatures beneath archaeological campfires using carbon stable isotope composition of soil organic matter. *Journal of Archaeological Science* 34 (6): 850–57.

Wheeler, P. E. 1991. The thermoregulatory advantages of hominid bipedalism in open equatorial environments: The contribution of increased convective heat loss and cutaneous evaporative cooling. *Journal of Human Evolution* 21:107–15.

Whelan, R. J. 1995. *The ecology of fire.* Cambridge: Cambridge University Press.

Whitcome, K. K., L. J. Shapiro, and D. E. Lieberman. 2007. Fetal load and the evolution of lumbar lordosis in bipedal hominins. *Nature* 450:1075–78.

Whitehouse, N. J. 2000. Forest fires and insects: Palaeoentomological research from a subfossil burnt forest. *Palaeogeography, Palaeoclimatology, Palaeoecology* 164:231–46.

Winter, D., N. Abolmaali et al. (2003). The endoscopically and radiological visualization of the human vomeronasal organ. *Otolaryngology—Head and Neck Surgery* 129 (2): 147–48.

Wittinger, L., and J. L. Sunderland-Groves. 2007. Tool use during display behavior in wild cross river gorillas. *American Journal of Primatology* 69 (11): 1307–11.

Won, Yong-Jin, and Jody Hey. 2005. Divergence population genetics of chimpanzees. *Molecular Biology and Evolution* 22 (2): 297–307.

Wood, Bernard, and Brian G. Richmond. 2000. Human evolution: Taxonomy and paleobiology. *Journal of Anatomy* 196:19–60.

Worthman, C. M., and M. Melby. 2002. Toward a comparative developmental ecology of human sleep. *Adolescent sleep patterns: Biological, social, and psychological influences*, ed. M. A. Carskadon. New York: Cambridge University Press.

Wrangham, Richard W. 2001. Out of the Pan, into the fire: How our ancestors' evolution depended on what they ate. In *Tree of origin: What primate behavior can tell us about human social evolution*, ed. F. B. M. de Waal. Cambridge, MA: Harvard University Press.

Wrangham, Richard, and N. L. Conklin-Brittain. 2003. Review: "Cooking as a biological trait." *Comparative Biochemistry and Physiology Part A* 136:35–46.

Wrangham, R. W., J. H. Jones, G. Laden, D. Pilbeam, and N. L. Conklin-Brittain. 1999. The raw and the stolen: Cooking and the ecology of human origins. *Current Anthropology* 40:567–94.

Wrangham, R. W., C. W. McGrew, Frans B. M. de Waal, and P. Heltne, eds. 1994. *Chimpanzee cultures*. Cambridge, MA: Harvard University Press.

Yamagiwa, J., T. Yumoto, M. Ndunda, and T. Maruhashi. 1988. Evidence of tool-use by chimpanzees (*Pan troglodytes schweinfurthii*) for digging out a bee-nest in the Kahuzi-Biega National Park, Zaïre. *Primates* 29:405–11.

Young, Larry J., and Zuoxin Wang. 2004. The neurobiology of pair bonding. *Nature Neuroscience* 7:1048–54.

Youngson, N. A., and E. Whitelaw. 2008. Transgenerational epigenetic effects. *Annual Review of Genomics and Human Genetics* 9:233–57.

Zeitzer, Jamie M., Derk-Jan Dijk, Richard E. Kronauer, Emery N. Brown, and Charles A. Czeisler. 2000. Sensitivity of the human circadian pacemaker to nocturnal light: Melatonin phase resetting and suppression. *Journal of Physiology* 526 (3): 695–702.

Zeller, A. 1992. Communication in the social unit. In *Social processes and mental abilities in nonhuman primates*, ed. F. D. Burton. Queenston: Edwin Mellon Press.

Zihlman, Adrienne. 1996. Reconstructions reconsidered: Chimpanzee models and human evolution. In *Great ape societies*, ed. W. McGrew, L. Marchant, and T. Nishida. New York: Cambridge University Press.

Zihlman, Adrienne, Deborah Bolter, and Christophe Boesch. 2004. Wild chimpanzee dentition and its implications for assessing life history in immature hominin fossils. *Proceedings of the National Academy of Sciences* 101 (29): 10541–43.

Index

The letter *t* following a page number refers to a table on that page. Page numbers in *italic* type refer to a figure.

critical day length (CD), breeding and, 54

cryptochromes, 47–48, 52

culture: cognitive abilities and, 134–35; Kroeber on, 131; primates and, 12–13, 109, 114–15, 131–39, 174–75; speech and, 133–34, 135–39

Czeisler, C. A., 53

Dainton, M., 94

Dart, Raymond, 153, 154, 155

Darwin, Charles, 12

de Lumley, H., 168

divergence of chimpanzees and humans, 6, 69, 72–75; ASPM and, 77; atmospheric carbon dioxide and, 78–79; bipedalism and, 7–8, 173–74; brains and, 84–86; C4 and, 80–81; Chad fossils and, 99–103; dating of, 82, 173; domestication and, 74; environmental sources of, 77–81; fossil evidence of, 99–105; gene duplication and, 75; gene number and, 76–77; Great Rift Valley and, 77–78, 79; growth rates and, 86–87; HAR family genes and, 76; indels and, 75; Kenya fossils and, 99; reverse transcription and, 76; selection for cognitive development and, 75–77; speech and, 86; synaptic branching and, 77; tandem repeat sequences and, 74; teeth and, 87

divergence of gorillas and chimpanzees, 87–88

DNA, epigenetic influence and, 66–67

Dordogne, France, 169

Efe Pygmy, 146

endocrinology, 65

end-state comfort effect, 159

entrainment: circadian rhythms and, 60–61, 64; flexibility of, 65–66; process of, 180

epigenetics, 64, 66

epochs and their dates, 79t

Evans-Pritchard, E. E., 3

evolution: adaptive radiation and, 89–90, 101–2; clades and, 90; domestication and, 74, 87; gene duplication and, 75; genetic drift and, 74; human, charting the course of, 89–90; linear model of, 101; punctuated equilibrium and, 73. *See also* primates, evolution of

eye, human, 50–56; apoproteins and, 51; fovea and, 51; mediation of the visible spectrum and, 50–56; photopigments and, 51; photoreceptors and, 50–51

Falk, Dean, 97, 153

fear: fire and, 31–34, 176; image and emotion and, 31; serotonin and, 31

fire, association with, 171–72; acquisition model and, 43, 176; Baganda people and, 186; behavioral prerequisites and, 4–5; benefits of, 15–16, 23–24, 28; bipedalism and, 5; birds and, 23–24; categorization and nurturing of, 41–42; cognitive development and, 29–30; evidence of, 141–47; extinction and, 30–31; fear and, 32–33, 176; food resources and, 24; functions of, 144–47;

grooming behavior, 5, 11, 112, 115, 118–20, 121; as currency, 121; as social adhesive, 121, 123, 128

Guayaki people, 186

Haile-Selassie, Yohannes, 104

Harlow, Harry, 132–33

Harris, J., 42, 154–55

hearths: archaeological methods and, 147–63; versus campfires, 15; evidence of, 15, 141–43; *Homo erectus* and, 143; Middle Pleistocene and, 142–43; Swartkrans and, 143

histones, 66

home base: as emergent behavior, 164; social and cognitive implications of, 165

hominins: the Ancestor and, 90–91; bipedality and, 69; cave habitation and, 144; definition of, 12; domestication of fire and, 18; evolution, speed of, and, 172–73

hominization, 105–7; *Homo erectus* and, 144; self-domestication and, 107

Homo, 6; dating of, 104–5; tandem repeat sequences and, 106–7

Homo erectus, 152, 156; big-game hunting and, 17; brain capacity of, 98; in China, 167–68; China and, 166–68; cosmopolitanism and, 166; domestication and, 106; fire and, 141, 143; hearing and, 137–38; hominization and, 144; roasting tubers and, 16; teeth and, 87, 143; use of fire and, 166

Homo ergaster, 152; big-game hunting and, 17; domestication and, 106

Homo habilis, 98, 106

Homo rudolfensis, 106

Homo sapiens, 188; domestication of fire and, 9

Homo sapiens sapiens, 71, 158

hormonal triggers, reproductive sequence and, 56–61; gonadotropin-releasing hormone (GnRH) and, 57–58

Hula Valley, Israel, 165–66

Human Genome Project, 72, 75

Human Relations Area Files, 185

hypothalamus, suprachiasmatic nucleus of, 46

Igbo people, food roasting and, 145–46

illuminance: luminous intensity and, 61–65; values, lux and, 60t

innovation: the Ancestor and, 140; traditions and, 115–16, 117; youth and, 13–14, 113–14, 176

insects, 10; fire and, 7, 16, 21–22, 35, 172, 176; nutritional value of, 25–27. *See also* pyrophilus insects; termites

isoflavones, 56

Jacobsen's organ, 53–54

Jacobsen's organ (vomeronasal), 53–54

Jirtle, R. L., 66–67

Kibale Forest, Uganda, 55

knuckle-walking, 86, 88, 92–94, 158, 159, 173

Koobi Fora, Kenya, 13–14, 18, 70, 106, 160–62, 169; controlled fire and, 144–45

Kroeber, Alfred, 135; on culture, 131
Kromdraai, South Africa, 151–52
Kummer, Hans, 110

Laetoli, Tanzania, 38
Lahn, Bruce, 172
Lamb, Charles: "On Eating Roast Pig", 17
lateral geniculate nucleus (LGN), 52
Lazaret, France, 169
Leakey, Mary and Louis, 154
Leonard, William R., 98
leptin, 56–57
Liebermann, Philip, 135
light-dark cycles: hormonal cycles and, 10; language acquisition and, 11; social structure and, 10
light intensity: emission versus reception of, 61–65; mammals and, 54–55; plants and, 55–56; prosimians and monkeys and, 55
Linnaean rules, 99
Livingstone, David, 23
Lokalalei, Kenya, 159–60
Lonergan, N., 12
lumbar lordosis, 92–93, 173
lux, 50; values, illuminance and, 60t

Macaca sylvanus, 133–34. *See also* Barbary apes; Gibraltar monkeys
macaques: babysitting and, 113; bipedalism and, 92–93, 95–96; hierarchy and, 111; insects and, 25; social behavior and, 116; socialization of the young and, 118. *See also* primates, nonhuman

Makapansgat, South Africa, 151–52; osteodontokeratic culture and, 154
Masai people, 42
Mayr, Ernst, 41
McGrew, Bill, 26, 33, 121–22, 128, 131
meat, primates and, 16–17
meiosis, 113
melanopsin, 47, 53
melatonin: as antioxidant, 48–49; circadian rhythms and, 47, 48–49, 180–81; light and, 61, 64–65, 67, 187–88; memory formation and associative patterns and, 11; pituitary and, 49; plants and, 55–56, 183; progression of seasons and, 54; puberty and, 58, 59; sleep and, 59–60; synthesis of, 48
Melby, M., 15
methylation, 66–67
mind, 4, 10, 12–13, 18, 77, 132, 173, 177, 188; consciousness and, 133; culture and, 12–13; fire as an adaptive tool and, 13–14; prosimians and, 12–13; theory of, 28, 130
Miocene, 19–20, 69–70; apes and, 12; bipedality and, 69–70; contemporary ape forms and, 89; seasonality and, 182
Miocene, Early: climate changes and, 178; habitats and, 79–81; mammalian faunas and, 79–81
Miocene, Late, 79–81, 178, 182; climate changes and, 24; nonhuman primates and, 107; use of fire and, 69–70
mirror neurons, 136
Mode 1, 2 tools. *See under* tools

monkeys: babysitting and, 112, 118; behavior of, 11–12, 111; birth of, 118; brain and, 86, 136; caves and, 144; communication and, 133, 138; food behavior and, 16; geophagy and, 24; grandmothering and, 112–13; hands and, 83–84, 90; hierarchy and, 111–12; information processing and, 114–15, 118, 125–27; insects and, 9, 25–26, 35; meat and, 16; memory and, 24, 85; reproduction and, 54–55; self-medication and, 104; theory of mind and, 28, 130; tools and, 70, 147; tradition and, 116–17, 125. *See also* Gibraltar monkeys; primates, nonhuman

Nakali, Kenya, 82
Neanderthals, 84–85
New Guinea, 41
nucleosomes, 66
Nuer people, 42

Olduvai Gorge, East Africa, 168
Omo I and II, Kenya, 71
"On Eating Roast Pig" (Lamb), 17
opioids, 31
Orrorin tugenensis, 28, 102–3, 175–76, 178; habitats of, 105

Paabo, Svante, 98
Pan, 12, 101, 103
Pang, S. F., 183
Pan paniscus, 122
Pan troglodytes, 122
Pei, Wenshong, 167
Peking Man (Shapiro), 167

Pfeiffer, John, 67
phase shift, 67; circadian rhythms and, 45, 47, 67; firelight and, 67; natural selection and, 61
photometry, 49–50
photons, transduction and, 51–52
photoperiodicity, 46–47
photoreception: history of, 53; human eye and, 50–51
phytoestrogens, 56
Piaget, Jean, 125
Pierolapithecus catalaunicus, 88; wrists of, 94
pineal gland, 52
pituitary, melatonin and, 49
plants: CO_2 and, 78–79; fire and, 21, 37; as medicine, 104, 149; melatonin and, 55–56, 183; photosynthesis and, 36; phytoestrogens and, 183; toxicity and, 24, 178
Pleistocene, Middle: hearths and, 142–43; home base and, 165
Pliocene, 81
point mutations, 72–73, 75
Pomo and Miwok peoples, 145
primates: color perception and, 53–54; hormonal triggers, reproductive sequence and, 56–61; meat and, 16–17; mind and culture and, 12–13; transduction and, 52
primates, evolution of, 81–88; brains and, 84–85; evolutionary body reorientation and, 83–84; hands and, 83–84; social groups and, 110–21; tropical habitats and, 82; vertical orientation and, 82–83

species germinalis (initiating or germinating species), 97, 101
speech: anatomical requisites of, 135–36; culture and, 133–34, 135–39; generative ability and, 138–39; genes and, 137–38; neuronal precursors of, 136–37; thoughts and, 138; vocal-auditory channel and, 137–38
Sponheimer, Matt, 156–57
Sterkfontein, South Africa, 151–52
Struhsaker, Tom, 138
Strum, Shirley, 117
suprachiasmatic nucleus, 52, 60t
Swartkrans, South Africa, 13–14, 143, 151–52, 153, 162–63; dating of, 155–56

tandem repeat sequences, 73–74; *Homo* and, 106–7
Taung, South Africa, 151, 152–54, 177
termites: African savanna and, 22; chimps and, 124; nutritional value of, 16; tools and, 7, 9, 26, 35, 128–29, 156
Thomas, Elizabeth Marshall, 40–41
time giver (Zeitgeber), 47, 64, 67
Tomasello, Michael, 134–35
tools: apes and, 7; bonobos and, 8; chimps and, 35, 128–29, 149, 156, 158; conservatism and, 144; fire and, 6–7; hominization and, 105–6; materials and, 38, 40; Mode 1, 144, 154, 157, 159; Mode 2, 70, 144, 147, 158, 162;

monkeys and, 70, 147; nonhuman primates and, 158–59; termites and, 7, 9, 26, 35, 128–29, 156. *See also specific archaeological sites*
tradition, 117
transduction: melatonin and, 48; photons and, 51–52; primates and, 52

Upper Paleolithic: fire and, 142, 145; tools and, 158

visible spectrum: eye's mediation of, 50–56; psychophysical properties of, 49–50
vitamin B$_{12}$, 26, 53, 66, 178–79
vitamin C, 80, 179, 187
vomeronasal, 53–54
vomeronasal (Jacobsen's organ), 53–54

Waterland, R. A., 66–67
wood, 38–40
Wood, Bernard, 12
Worthman, C. M., 15

Xihoudu, China, 168

Yaghan people, 96

Zhoukoudian, China, 167–68
Zihlman, Adrienne, 87, 122